# Lecture Notes in Economics and Mathematical Systems

**437**

**Springer**
*Berlin*
*Heidelberg*
*New York*
*Barcelona*
*Budapest*
*Hong Kong*
*London*
*Milan*
*Paris*
*Santa Clara*
*Singapore*
*Tokyo*

Carsten Jordan

# Batching
# and Scheduling

## Models and Methods
## for Several Problem Classes

 Springer

Author

Dr. Carsten Jordan
Christian-Albrechts-Universität zu Kiel
Institut für Betriebswirtschaftslehre
D-24089 Kiel, Germany

Library of Congress Cataloging-in-Publication Data

Jordan, Carsten, 1965-
   Batching and scheduling : models and methods for several problem
classes / Carsten Jordan.
      p.   cm. -- (Lecture notes in economics and mathematical
systems ; 437)
   Includes bibliographical references and index.
   ISBN-13: 978-3-540-61114-1      e-ISBN-13: 978-3-642-48403-2
   DOI: 10.1007/978-3-642-48403-2
   1. Scheduling (Management)  2. Economic lot size.   I. Title.
II. Series.
TS157.5.J67   1996
658.5'3--dc20                                           96-15317
                                                           CIP

© Springer-Verlag Berlin Heidelberg 1996

Softcover reprint of the hardcover 1st edition 1996

Typesetting: Camera ready by author
SPIN: 10516249       42/3140-543210 - Printed on acid-free paper

# Acknowledgments

This work is the result of my research performed at the Christian-Albrechts-Universität zu Kiel under the supervision of Professor Dr. Andreas Drexl. I am very grateful to him for creating a stimulating research environment and for his guidance, support and motivation during the preparation of this work. I would also like to thank Professor Dr. Klaus Brockhoff for undertaking the task of the secondary assessment.

I am indebted to my colleagues at the Institut für Betriebswirtschaftslehre, especially to Dr. Knut Haase, Sönke Hartmann, Alf Kimms, Dr. Rainer Kolisch, Andreas Schirmer and Dr. Arno Sprecher for many suggestions and helpful discussions; Andreas proof-read the entire manuscript.

Furthermore, I would like to express my gratitude to Dr. W. Brüggemann, Prof. Dr. D. Cattrysse, Prof. Dr. B. Fleischmann, Prof. Dr. M. Salomon and Dr. M. Solomon for the help they provided with papers, computer programs and test instances.

For the case description I am indebted to Dr. Jürgen Böttcher from the BEIERSDORF AG.

With respect to the technical realization, I owe thanks to Justus Kurth for his suggestions and the programming, and to Uwe Penke for the technical support in the computer laboratory.

My wife deserves thanks for her support throughout the whole work.

Carsten Jordan

# Contents

# Chapter 1

# Introduction

In some manufacturing systems significant setups are required to change production from one type of products to another. The setups render the manufacturing system inflexible as for reacting to changes in demand patterns, hence inventories must be maintained to guarantee an acceptable customer service. In this environment, production scheduling faces a number of problems, and this work deals with mathematical models to support the scheduling decisions.

Some more background and motivation is given in the following sections, as well as in a case description in Section 1.3. The synopsis in Section 1.4 outlines the topics of the work.

## 1.1 Motivation of the Planning Problem

Consider the production of metal sheets in a rolling mill. If the width of the next type of sheets is greater than the width of the preceding type, then the roll needs a setup: during the rolling process the edges of a sheet cause grooves on the rolls' surface, thus, the surface must be polished if a greater width is run next. Sheets with a smaller width can be run directly, without a setup. Another example in which setups are sequence dependent is a line where cars are sprayed: if the color of the paint changes, the cleaning of the tools requires a setup depending on the sequence of the colors. Only a small setup may be needed for changing from a light to a dark color, but a thorough cleaning of the tools is required for a setup from a dark to a light color. In the case description in Section 1.3 we describe a more complex example. Setups occur not only in manufacturing systems but in computer systems as well. Consider different tasks to be executed on one processor, where each task requires a certain compiler to be resident in memory, and some of the tasks share the same compiler. A setup in this system occurs if the compiler must be loaded before the execution of the task.

We will call the (manufacturing) facility *processor* or *machine*, and, depending on the level of aggregation, a machine may represent one specific machine, a line or a whole factory. If we consider only one facility we deal with the *single-machine case*, otherwise with the *parallel* or *multi-machine case*. A machine processes or produces different types of products, referred to as *families* or *items*, and setups are required in between these families. Synonym names for families are classes or (product) types. We will say production or processing even if the real action on a machine is, for instance, a test or assembly operation. Production on a machine serves to fulfill (exogenous) demand. At the rolling mill for instance, orders for different types of metal sheets may be due at some due date, or cars painted in different colors must be ready for assembly before delivery time. Demand may also be internal if orders must be delivered in time for downstream operations. The quantity of a family or an order (belonging to a family) to be delivered at a certain point in time, is a *job*, each job belonging to one family. We are primarily concerned with problems with a large number of jobs but a small number of families. The time by which a job must be delivered is called a *deadline*.

Production scheduling now faces two types of constraints:

Due to the setups and the capacity restrictions of the manufacturing system, the production and the demand pattern will typically be non-synchronized. This non-synchronization results in either inventory or back-orders. In order to be competitive (or to fit into a larger "framework"), the degree of non-synchronization between production and demand pattern must not be "too big". E.g. the company wants to guarantee the customers a certain service level, and/or inventory must be kept below a certain level. These are *service* constraints: they reflect economical considerations concerning the market position or the capital tied up in inventory.

The other type of constraints are *materials flow transformation* constraints[1] (related to the production process), which in our case means that setups occur between families.[2] Setups depend on the technological characteristics of the machine. A setup may for instance involve the preparation, cleaning or heating of a machine *before* production can start; after the setup, the machine is in a certain setup state and a certain family can be produced. We assume that the setup state does not change during processing[3], and that normally setups incur setup times and setup costs. A synonym for setup is changeover. Setups consume the limited capacity of the machine without contributing to production. We can omit a setup if two jobs of the *same* family are produced in one run, i.e., both jobs are in one *batch*. *Batching* thus means to group jobs of one family together. From an economical viewpoint this provides one way to realize economies of scale.

The fundamental planning problem that is the focus of this work deals with the balancing of production against demand in the presence of both materials flow transformation and service constraints, in order to achieve a "good" schedule. We refer to this problem as the *batch sequencing problem* (BSP). The name characterizes the interdependent decisions that are to be made: jobs must be batched and sequenced. It is necessary to find a compromise here, because in general we can neither batch all jobs of one family nor produce "lot for lot", i.e. produce jobs in order of increasing deadlines. In the former case, we would violate deadlines and/or keep high inventories, in the latter case too much time would be wasted in setups.

In the mathematical models in the next chapters we assume that demand must be fulfilled without back-logging, i.e. that jobs have a certain processing time and an associated deadline. Jobs may not take longer than the deadline to be completed. Completing a job earlier than its deadline incurs holding costs while the setups incur setup costs. In determining a good schedule we concentrate on *cost* minimization of the *production* process. The costs of inventories and setups are difficult to quantify, as they rarely reflect real out–of–pocket

---

[1]Cf. also Kuik et al. [77] for the distinction between service and materials flow transformation constraints.

[2]Moreover, we consider multi-level product structures as materials flow transformation constraints.

[3]An example in which production is a heat treatment and this assumption does not hold (the setup state changes during production) is described in Pinedo [99].

costs (while setup times can be measured easily in most cases). However, holding and setup costs express the preference of the decision maker in the mathematical models to avoid both inventories and setups. In other approaches deadlines may be violated (one considers *due dates* instead of deadlines); in this case, often the minimization of maximum lateness is considered as objective. Maximum lateness is used as an equivalent for the loss of goodwill of customers; this approach focuses on the *market* rather than on the production process.

## 1.2   Background and Related Work

A solution of the BSP will always be a compromise in which neither inventories nor setups can be avoided. Therefore, in practice, sometimes efforts are not directed to *solve* the BSP but rather to *eliminate* it: for instance, one way to achieve zero inventory is just-in-time (JIT) production, where the production is is entirely driven by demand. For JIT production, there should be only a small number of products with a relatively constant demand. The planning problem of when to produce and how much then disappears, but, the manufacturing system itself must be suited for JIT production, i.e. must be "flexible" and setups must not be significant. The technological developments of the last decades have resulted in major improvements in this direction. Today, batch production systems with large setups can often be replaced by flexible manufacturing systems, leading to improved customer service and reduced inventories. However, in general this is done at the expense of higher investments.[4]

There are nonetheless many manufacturing systems where a JIT production is impossible, as in the above examples with significant setups. In other problem settings, demand may be highly seasonal and/or it is cheaper to hold inventory than to maintain slack capacity. If, for example, investments for the manufacturing system are high and the value of finished goods is low, holding inventory may be a reasonable option.

As the BSP concentrates on short term planning, it is only part of an overall production

---

[4]Haase and Göpfert [64] describe a case where high investments for capacity expansion are avoided by means of an improved planning process such that demand can be satisfied with the existing facilities.

planning and control system in which setups and capacity restrictions must be taken into account (cf. Drexl et al. [41]). We simplify the planning problem in the sense that data are assumed to be deterministic and to be given or forecasted for a certain planning horizon, e.g. the orders for the rolling mill are assumed to be known in the next three months. To cope with the uncertainty of future data, the BSP is solved within a rolling horizon environment:[5] a plan is made for the entire horizon, but only the first part of the plan is implemented. Data are then updated and planning is repeated. Thus, the BSP must be repeatedly solved as a subproblem for decision support in short term planning. This is the major motivation to develop fast algorithms for the BSP.

Finding a good compromise between the number of setups and the inventory level is not a new problem and neither is its treatment in literature. The economic order quantity of Andler or Harris is the first attempt in this field; for a historical review cf. Erlenkotter [48]. We briefly review some of the literature and outline the way in which this approaches differ from the BSP. An extensive up-to-date overview of batching problems is presented in Kuik et al. [77].

A dynamic version of the economic lotsize problem is the well-known model of Wagner and Whitin [123], which is now solvable in linear time, cf. e.g. Domschke et al. [39]. Wagner and Whitin examine the uncapacitated one-family case, but some of their insights are helpful for the BSP as well. A multi-family model, having, however, a constant demand rate, is the economic lot scheduling problem (ELSP) which remains difficult to solve. Dobson [38] presents algorithms for a problem with sequence dependent setups. In the capacitated lotsizing problem (CLSP) we do not decide about the timing of jobs but about the size of lots in each period. The CLSP is based on the assumption that a setup is incurred for each lot in a period. Demand is given for (many) products and (few) periods. Periods are large and the CLSP solves a more aggregate problem than the BSP. An approach based on lagrangean relaxation is presented by Diaby et al. [36] and Tempelmeier and Derstroff [119]. Diaby et al. consider the single-level case and solve problems with up to 5000 products and 30 periods; Tempelmeier and Derstroff consider

---

[5]Ten Kate [74] embeds the BSP in the context of *order acceptance*. Domschke et al. [39] describe how lotsizing problems are solved in a rolling horizon environment.

multi-level structures. In the basic CLSP, problems of scheduling are not dealt with (so that sequence dependent setups cannot be considered), which has motivated researchers to consider lotsizing *and* scheduling problems. One example of this is the discrete lotsizing and scheduling problem (DLSP) (cf. Fleischmann [49]) which is in fact very similar to the BSP: Chapter 4 is devoted to a comparison between DLSP and BSP. Other lotsizing models that deal with scheduling aspects are the proportional lotsizing and scheduling problem (PLSP) (cf. Haase [61]), or a CLSP variant in which lots may be linked over periods (cf. Haase [62] and Haase and Kimms [63]).

## 1.3   Case Description

The following case description illustrates an industrial background for the theoretical models presented in the following chapters.

The BEIERSDORF AG is a manufacturer of branded goods with three lines of business, **Tesa** for adhesive tapes, **Cosmed** for cosmetic, and **Medica** for medical products. The production facility for the **Cosmed** line of business, e.g. body and suntan lotion or skin care cream, is located in Hamburg, Germany.

The production process is as follows: the lotion or cream (referred to as *filling*) is mixed/ produced according to its recipe in big containers/mixers and then filled in a tank. There are about 120 different fillings. The filling is then bottled from the tank on one of the 22 bottling lines, after which the tank must be empty. Production is make–to–stock, and the goods are then shipped from stock to the customers (which are commercial enterprises in most cases).

The bottling lines are the bottleneck in the production system and thus are scheduled first. On the bottling lines setups must be taken into account. The largest setup occurs if the bottle size is changed, which takes a time of up to one shift: many devices on the line must be changed and require new adjustment. Furthermore, for some bottle sizes, devices (called "Formatbecher") to position the bottles correctly on the line are necessary. If the bottle size changes, the "Formatbecher" must be changed, too.

Another (but smaller) setup on the line occurs if the filling is changed: the filling device of the line must then be cleaned, and the thoroughness of the cleaning depends on the sequence of fillings. E.g. suntan lotion with a *low* protection factor can follow a lotion with a *high* protection factor without cleaning, because a minimum protection factor must be guaranteed for the lotion (but the inverse is not true).

Only a small setup is required if the label on the bottle is changed: for each country a different label is used.

In general, one filling is assigned to one line (because e.g. the lines for skin care cream and sun lotion differ technologically), therefore the whole problem with 22 lines and 100 fillings decomposes into one machine problems with 3 to 15 different fillings per line. For fillings needed in large quantities machines may be dedicated for some time for that filling (and hence, scheduling is trivial for those machines).

The final product is a certain filling, in a certain bottle with a certain label, called *article*. Demand for each article is forecasted, and the sales department develops sales plans. The production planning and control (PPC)-system compares forecasted sales and inventory level for each article and proposes a replenishment order if the inventory level reaches a certain safety level ($(s, S)$-policy). The replenishment order from the PPC-system generates a due date as well. Contrary to a deadline, this due date may be exceeded, but there must be no stock-out of an article.

The similarity to the above BSP is that from the "bottling line viewpoint" (concerning the setups) various articles belong to the same family, e.g. articles with the same filling and bottle size, but different labels.

Replenishment orders for articles can be interpreted as jobs with deadlines, and two jobs on a line belong to the same family if no (or a "small") setup is required between them. More precisely, in an *aggregation* step it is specified which articles form a family, and the problem is then modeled as a BSP: again, the planning problem is how to batch jobs of one family, and how to sequence the batches. The bottling line "prefers" large batches of *one* family while demand constraints may dictate that in a certain time articles of *different* families become available. The planning should determine a "good" sequence on the line and complete jobs as close as possible to the due date. The problem for the moment is that the proposals of the PPC-system do not take the setups into account, and that the proposals require numerous adjustments "by hand".

As is often the case in industrial problems, many circumstances make it difficult to apply an algorithm directly, as data acquisition takes a long time and a lot of very particular constraints must be taken into account. E.g., the production and the bottling of the fillings require synchronization, and planning cannot be done independently of the bottle supplier's capacity constraints. The first step, currently undertaken, to improve the planning is to integrate the sequencing planning TDS1 of the PPC-system R/2 of SAP[6], so that the PPC-system generates "more realistic" proposals.

## 1.4   Chapter Synopsis

This work covers three topics: ($i$). We analyze the BSP and some of its variants and extensions, and we develop exact and heuristic solution procedures. ($ii$). We state equivalence of the BSP with the DLSP. On that account, two questions arise: we examine, which one of the models is better "suited" for the underlying planning problem, and we compare the performance of solution procedures for the DLSP and the BSP. The equivalence between BSP and DLSP strongly motivates and validates the analysis and the development of algorithms. ($iii$). The BSP is solved with different general solvers, a mixed-integer-programming and two constraint programming solvers. Again, two aspects are of interest: how must we encode the BSP for the particular solver, and what is the solver's performance when applied to the same problem.

The work is divided into 6 Chapters:

**Chapter 1** provides an introduction to the BSP, as well as a case description which motivates the BSP within an industry-context.

In **Chapter 2** we analyze batching in more detail and define *batching types*. Formal descriptions of the BSP and its variants and extensions are then presented in order to provide some insight into how they differ from each other. A complexity analysis and the

---

[6]R/2 and R/3 are PPC-systems from SAP AG, Walldorf, Germany. TDS1 is a module in the R/2 system which takes sequencing into account.

derivation of structural properties of the BSP follow. The formal treatment allows clear statements to be made about the structural properties of the BSP used in the design of algorithms in Chapters 3 and 5. At the end of Chapter 2 we present an instance generator.

In **Chapter 3** we develop algorithms for the single-machine case based on the analysis in Chapter 2. We present two exact algorithms, a branch and bound and a dynamic programming algorithm, as well as heuristics, a simple construction and improvement heuristic and a genetic algorithm. All algorithms tackle the variants of the BSP in different ways based on the analysis in Chapter 2, and computational experiments are presented to assess their performance.

**Chapter 4** is devoted to a comparison between DLSP and BSP. We illustrate and formally state the equivalence between both models and then compare the solution procedures for the DLSP reported in literature with procedures developed in Chapter 3. The central clue in this chapter is that decision variables in the DLSP and BSP differ, but that the solutions in both models correspond to each other.

**Chapter 5** provides an algorithmic analysis of two extensions of the BSP: multi-level product structures and parallel machines. The algorithmic treatment is based on the analysis in Chapter 2, and restricted computational results offer insight into the algorithmic performance.

In **Chapter 6** a comparison of different general solvers for the BSP is presented. We provide models for a mixed-integer programming solver as well as for two constraint programming solvers. Using the BSP as the example problem, we compare the encoding for the different solvers, i.e. the question of "how to represent the problem for the computer", as well as the performance of the solvers for a small set of instances.

Finally, **Chapter 7** summarizes the work and suggests on directions for future research.

# Chapter 2

# Description of the Problems

This chapter is devoted to a formal description and analysis of the Batch Sequencing Problem (BSP). We review the literature of batching and scheduling and define batching types in Section 2.1. The notation and a three-field descriptor for the BSP follow in Section 2.2; the descriptor will allow us to refer to the variants of the BSP in a short and concise manner. Formal models as well as an example for the single-machine case are presented in Section 2.3, multi-level and parallel machine models and examples in Section 2.4. We examine the complexity of the problems in Section 2.5, where it turns out that the BSP with nonzero setup times is NP–hard. Therefore, an analysis of structural properties in Section 2.6 is warranted. The results of Section 2.6 and the instance generator described in Section 2.7 are used when we present algorithms in Chapters 3 and 5.

## 2.1   Batching Types and Literature Review

Generally, we can distinguish between scheduling *with* and *without* batching. The largest part of the scheduling literature addresses the latter case; problems are concerned with e.g. precedence and/or resource constraints, and even without setups difficult problems arise.

For batching *and* scheduling there is not much literature for a number of reasons: first, if batching addresses issues related to sequence dependency, problems are often dealt with in

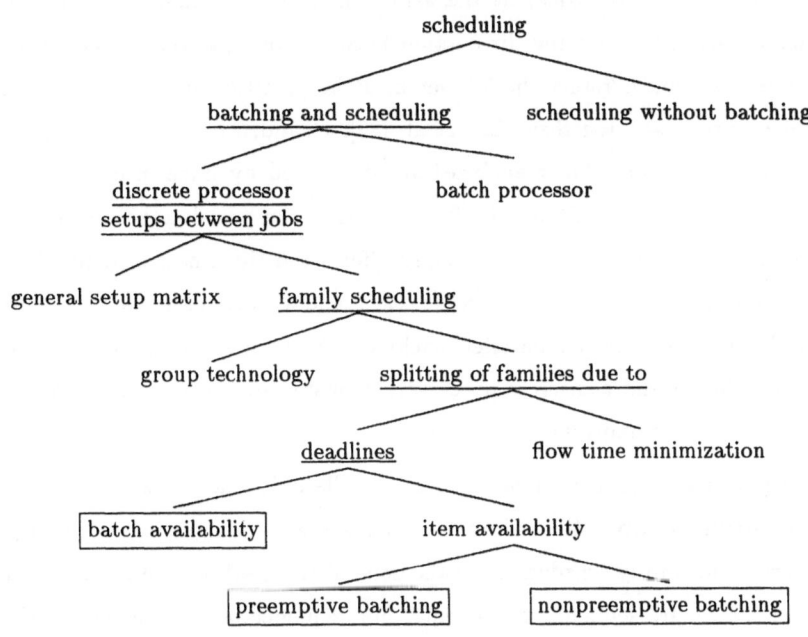

**Figure 2.1:** *Classification of Batching and Scheduling*

the context of the traveling salesman problem (TSP). Second, batching is often identified as lotsizing, so research concentrated in this area. Third, for batching and scheduling there exists no "standard problem", such as e.g. the job shop, TSP or the resource constrained project scheduling problem. Consequently, the terminology is not homogeneous and it is difficult to compare different approaches.

In order to place this work in the context of batching and scheduling we use the classification depicted in Figure 2.1, in which the characteristics of the BSP are underlined. We must distinguish between different *batching types* (in boxes in Figure 2.1) as outlined in the following.

In batching and scheduling, the machine or the processor may either be a discrete processor or a batch processor. A **batch processor** processes several jobs simultaneously. A simple example for a batch processor is an oven. Cookies represent jobs, and the oven processes several jobs (=cookies) at the same time. In scheduling a batch processor the primary constraint is not the disjunction between jobs (so that a sequence of jobs would have to be found), rather the job set must be partitioned to decide which of the jobs should be processed together. Lee et al. [83], Chandru et al. [26] and [27] analyze batch processing systems. Their analyzes are motivated by burn-in operations for the testing of semiconductor circuit boards. The testing of the boards takes place just before delivery, so that the testing schedule strongly affects the timeliness of deliveries. Fairly simple polynomial algorithms exist for some cases, and a worst case analysis is presented for more difficult cases, e.g. for parallel machines. Ahmadi et al. [3] consider flow-shop systems with a batch and a discrete processor. Webster and Baker [124] summarize the results for some polynomially solvable cases.

A **discrete processor** is a machine that can handle only one job at a time, and there are **setups between jobs**. On a discrete processor we distinguish between a **general setup matrix** and family scheduling. In the general case, setups are given for each pair of jobs,[1] and setups are sequence dependent.[2] For sequence dependent setups, makespan minimization is equivalent to the TSP, cf. Conway et al. [32] and Baker [8]. Therefore, in scheduling with general setups, additional constraints and/or different objectives are considered. Motivated by the schedule of observations for an orbit telescope, Moore [94] proposes a mixed-integer programming formulation where the number of tasks scheduled between their release dates and deadlines is maximized. Barnes and Vanston [10], Laguna et al. [79] and Bernardo and Lin [13] consider objectives, in which the sum of setup *and* tardiness costs of jobs must be minimized. Makespan minimization with jobs subject to release dates is the objective in Bianco et al. [15]. Ovacic and Uzsoy [97] minimize maximum lateness and propose a rolling horizon algorithm, where the problem is solved partially within the horizon with an enumerative algorithm. In Ovacic and Uzsoy [96], approximation algorithms for parallel machines and makespan minimization are presented. Finally, Rubin and Ragatz [107] present a genetic algorithm for the total tardiness problem with general setups.

---

[1]If setups are *tool switches* of a tool magazine, setups may be more complicated, cf. Locket and Muhlmann [88].

[2]In the sequence independent case, setups can be associated with each job and batching is eliminated.

Often, the general setup matrix has a special structure such that the jobs can be partitioned into families, and setups occur between families. This is close to the problem described in Chapter 1 and is called **family scheduling**. Family scheduling is also closely related to lotsizing, cf. Chapter 4. Scheduling with general setups is as difficult as the TSP, but in family scheduling models, the family structure can be exploited.

A simplifying assumption in family scheduling is that the number of setups equals the number of families, called **group technology** assumption (cf. Potts and Van Wassenhove [102]). Under the group technology assumption problems can often be solved with two level approaches: first, within a family, and then (with families as composite jobs) on the family level.[3] Much more difficult yet more realistic problems arise if families are split into several batches and the number of setups for each family is part of the decision. The majority of the body of literature deals with flow time minimization, a much smaller part deals with deadlines for the individual jobs. In this context we consider different *batching types*, which determine when a job becomes available with respect to the batch completion time. We distinguish between:

- **item availability–preemptive batching** (*ia-pb*): each *item* in the batch, i.e. the jobs, becomes available at the completion time and idle time may preempt the batch.

- **item availability–nonpreemptive batching** (*ia-npb*): jobs become available at their completion times but jobs in one batch are processed without idle time between them. If there is idle time between consecutive jobs of the same family, a new setup is needed.

- **batch availability** (*ba*): jobs become available at the completion time of the batch, i.e. all jobs of a batch are available as soon as the last job in the batch completes.

Figure 2.2 illustrates the different batching types we consider in this work. The jobs $(1,1)$ and $(1,2)$ (family, job-number) are depicted at their deadlines. If both jobs form a batch,

---

[3]E.g., for the sum of completion times criterion, it is optimal to apply the shortest processing time (SPT) rule on both levels. First, jobs are ordered by SPT within the families, and then, with a sequence *independent* setup associated with each family, families are composite jobs. In the second step the families are ordered by SPT, giving the optimal schedule (cf. Webster and Baker [124]).

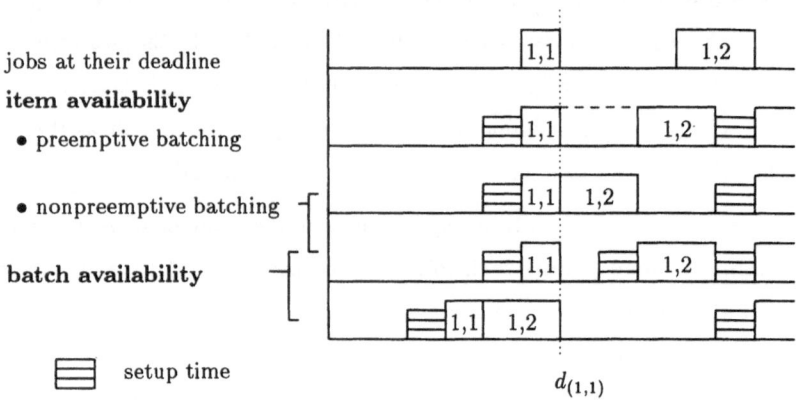

**Figure 2.2:** *Gantt Chart Illustration of Batching Types*

the batch is completed at the completion of job $(1, 2)$. In the item availability case, job $(1, 1)$ is available at its deadline $d_{(1,1)}$ even if the batch is completed later. In preemptive batching, we do not need a second setup to job $(1, 2)$ if there is idle time between both jobs. In nonpreemptive batching, job $(1, 2)$ must be leftshifted to avoid a second setup. Otherwise, both jobs form two batches. In batch availability, the batch must be completed before the deadline of job $(1, 1)$ if both jobs are batched, so that job $(1, 1)$ becomes available at its deadline. In batch availability, the last job in a batch must complete before the first job's deadline so that a batch does not contain idle time; hence, we always have nonpreemptive batching. For a batch processor, batch availability is implicit in its mode of operation.

As an example for item availability consider a production line which needs a setup before processing a certain type of products, but where the products become available after completion. If the setup state is kept when the line is idle, we have preemptive batching (e.g., because the devices on the line remain installed or only "small" changes must be made). If production requires continuous processing (as e.g. is the case for some pharmaceuticals), the mode of operation is nonpreemptive batching.

Probably the first work that distinguished between item and batch availability is Santos and Magazine [111], who state that problems tend to become more difficult in the latter case. Our distinguishing between preemptive and nonpreemptive batching, however, is new. Due to the fact that we consider deadlines *and* flow time of individual jobs, the question is how to insert idle time optimally making this distinction necessary.

An overview of what we call family scheduling is provided in Potts and Van Wassenhove [102], who also stress the relationship to lotsizing problems. Monma and Potts [92] present a general dynamic programming (DP) algorithm for different objective functions and examine the complexity of some family scheduling problems. The DP algorithm is polynomially bounded in the number of jobs, but exponential in the number of families. In Monma and Potts [93] heuristics for parallel machine scheduling are examined. Potts [101] presents a polynomial algorithm if there are only two families.

A number of sources consider the case in which families must be split due to the **minimization of flow time** with **batch availability**. Naddef and Santos [95] and Coffman et al. [30] show that a greedy algorithm is optimal if there is only one family and job processing times are equal. Then, the optimal number of batches is a function of the setup time, the number of jobs and the processing time. For the multi-family case [95] and [30] present approximation algorithms. Albers and Brucker [6] extend the analysis to weighted flow time and precedence constraints and delineate the borderline between hard and easy problems.

Other references consider the **minimization of flow time** with **item availability**. Vickson et al. [122] study a problem in which each job requires a component unique to that job and a component common to all jobs. Both components are processed on the same machine, and a setup is needed prior to the processing of the common component. They derive properties of the optimal solution and propose a DP algorithm to minimize flow time. The multi-family case is considered in Psaraftis [105], Ahn and Hyun [5] and Mason and Anderson [90] (in chronological order). Psaraftis' study is motivated by an aircraft landing problem: jobs are aircrafts waiting to land, families are the aircraft types, and the (sequence dependent) time distance between two aircraft types depends on safety regulations. E.g., a small-sized airplane may not land shortly after a wide-body aircraft, but the inverse order is possible. In [105] a DP algorithm is given for different objectives. Ahn and Hyun [5] extend this approach to unequal job processing times and propose an interchange heuristic. For sequence independent setups and weighted flow time, Mason and Anderson [90] identify properties of optimal schedules and derive dominance criteria and a lower bound for a branch and bound algorithm. Gupta [59] proposes a simple greedy heuristic for the above problem. Finally, in Ghosh [55] the multi-machine case with weighted flow time is considered. Ghosh shows that the problem is polynomially solvable for a fixed number of families, in which a (polynomially solvable) transportation problem appears as a subproblem.

In a number of sources deadlines or duedates for individual jobs are considered, but there are no setups. Posner [103] and [104] proposes algorithms for the deadline-constrained weighted completion time problem. Emmons [46] proposes dominance rules for a branch and bound algorithm to minimize mean flow time with the minimum number of tardy jobs. For the weighted tardiness problem with proportional weights, Szwarc and Liu [117] establish a partial ordering between jobs and are able to solve instances with up to 150 jobs. Bianco and Riciardelli [14] examine a problem with release dates for jobs (in which the time axis is reverted) and present a branch and bound algorithm with dominance rules to minimize total flow time. Hariri and Potts [65] minimize the weighted number of tardy jobs and solve problems with up to 300 jobs.

Literature regarding **family scheduling** that considers **splitting of families due to deadlines** is very sparse. So far, only **item availability–preemptive batching** has been dealt with. For sequence independent setups, Bruno and Downey [22] prove NP-hardness of the feasibility problem with nonzero setup times and of the optimization problem with nonzero setup costs. Unal and Kiran [121] consider only the feasibility problem, and they derive structural properties of solutions (cf. also Section 2.6). They propose a simple heuristic which is experimentally tested. Though the feasibility problem is NP–hard, the heuristic finds a solution for nearly all instances. The minimization of weighted flow time is considered by Park et al. [98]. They present a dominant set of schedules which is then used in a branch and bound algorithm. The minimization of total setup costs is considered by Hu et al. [68] and Driscoll and Emmons [44]. Hu et al. describe a polynomial algorithm for a special setup cost structure, Driscoll and Emmons propose a DP algorithm where dominance criteria are used to reduce the state space. Their study was motivated by a planning problem of a paper mill, which has sales commitments with deadlines. The sales commitments represent jobs, families are the types of paper sheets, and changing the types incurs setup costs. An optimal schedule minimizes the sum of setup costs while fulfilling all sales commitments in time. Sequence dependent setups are considered by Woodruff and Spearman [125]. They consider a problem where some of the jobs are "filler" jobs and may be left unscheduled. A tabu search algorithm is presented which maximizes the number of filler jobs scheduled.

Some references treat related or extended problems: Bednarzig [12] considers the job-shop problem with sequence dependent setups. Schutten et al. [112] consider release dates and due dates for jobs, and propose a branch and bound algorithm to minimize maximum lateness. In this problem some setups can be derived *a priori* from the time window associated with each job, and the main contribution is a tight lower bound based on the derived setups. In Ten Kate [74], setup as well as flow time is minimized, and simple construction and local search heuristics are proposed. Ten Kate embeds the BSP in the larger context of order acceptance and appears as a subproblem. Further references to problems related to batching and scheduling are Cheng et al. [28], Hochbaum and Landy [67] and Liao and Liao [86].

The literature review has demonstrated that the BSP with sequence dependent setups and different batching types has not yet been analyzed. Also, the (few) references up to now cover only the single-machine case. Finally, analysis of the BSP is motivated by the strong relationship to certain lotsizing and scheduling models; this is examined in more detail in Chapter 4.

## 2.2    Notation, Basic Definitions and $[α/β/γ]$ Descriptor

For notational purposes: we generally denote parameters with a small letter, decision variables with a capital letter, and sets with a calligraphic capital letter.

The BSP parameters are given in Table 2.1. Parameters related to the $N$ families are the family index $i$, the number of jobs $n_i$ in each family and the total number of jobs $J$. Index $i = 0$ denotes the idle machine. Holding costs $h_i$ denote the costs for holding one

**Table 2.1:** *BSP Parameters*

| | | |
|---|---|---|
| Related to the families | | |
| $i$ | : | index of the family, $i = 0, ..., N$, $i = 0$ : idle machine |
| $n_i$ | : | number of jobs in family $i$, $J = \sum_{i=1}^{N} n_i$ : total number of jobs |
| $st_{g,i}$ | : | setup time from family $g$ to family $i$ |
| $sc_{g,i}$ | : | setup costs from family $g$ to family $i$ |
| $h_i$ | : | holding cost for one unit of family $i$ |
| Jobattributes | | |
| $(i, j)$ | : | denotes the $j$-th job of family $i$, $i = 1, ..., N$, $j = 1, ..., n_i$ |
| $p_{(i,j)}$ | : | processing time of the $j$-th job of family $i$ |
| $d_{(i,j)}$ | : | deadline of the $j$-th job of family $i$ |
| $w_{(i,j)}$ | : | earliness weight of the $j$-th job of family $i$, $w_{(i,j)} = h_i p_{(i,j)}$ |

unit of family $i$ in inventory for one time unit. Setup times $st_{g,i}$ and setup costs $sc_{g,i}$ are given for each pair of families $g$ and $i$. The set of jobs is partitioned into mutually exclusive families $i$, and there are attributes for each job. The $j$-th job of family $i$ is indexed with a tuple $(i, j)$. A job $(i, j)$ has a processing time $p_{(i,j)}$, a deadline $d_{(i,j)}$ and a weight $w_{(i,j)}$. Job weights $w_{(i,j)}$ are proportional to the quantity (=processing time) of the job (*proportional weights*) and are derived from $h_i$ and $p_{(i,j)}$. We put the tuple in brackets to index the job attributes, because the tuple denotes a job as *one* entity (we have $j = 1, \ldots, n_i$, the index range of $j$ depends on $i$), whereas the indexes $g$ and $i$ for the setups denote a *pair* of families.

Setup times $st_{g,i}$, processing times $p_{(i,j)}$ and deadlines $d_{(i,j)}$ are given in time units ($TU$), setup costs $sc_{g,i}$ in monetary units ($MU$), and we assume all data to be integer valued. We assume the production speed ($U/TU$) to be equal to one, so that the processing time $p_{(i,j)}$ also denotes the number of units ($U$) produced for family $i$ in that time. The dimension of holding costs is $\left[ \frac{MU}{U \cdot TU} \right]$. We now can derive the dimension of $w_{(i,j)}$,

$$w_{(i,j)} \left[ \frac{MU}{TU} \right] = h_i \left[ \frac{MU}{U \cdot TU} \right] \cdot 1 \left[ \frac{U}{TU} \right] \cdot p_{(i,j)} \, [TU].$$

Inventory holding costs depend on the quantity and the type of products in inventory, and we use the weights $w_{(i,j)}$ to calculate the inventory holding costs. Therefore, $w_{(i,j)}$ is proportional to $p_{(i,j)}$ (=quantity) and $h_i$, the holding costs of the family $i$ (=type of products).

For the parameters, we make the following (not very restrictive) assumptions which free us from having to consider many exceptions in the sequel.

**Remark 2.1** *The following assumptions are made about the parameters of the BSP:*

1. *Setup times and setup costs satisfy the triangle inequality, i.e. $st_{g,i} \leq st_{g,l} + st_{l,i}$ and $sc_{g,i} \leq sc_{g,l} + sc_{l,i}$, $g, i, l = 0, \ldots, N$.*

2. *There are no setups within a family, i.e. $st_{i,i} = sc_{i,i} = 0$, and no tear-down times and costs, i.e. $st_{i,0} = sc_{i,0} = 0$, $i = 0, \ldots, N$.*

3. *"Longer" setup times lead to "higher" setup costs, i.e. $sc_{g,i} = f(st_{g,i})$ with a non-decreasing function $f(\cdot)$.*

4. *Jobs of one family are labeled in order of increasing deadlines, and deadlines do not interfere, i.e. $d_{(i,j)} + p_{(i,j+1)} < d_{(i,j+1)}$.[4]*

The assumptions in Remark 2.1 have several consequences. There are no setups between jobs of the same family, and the times and costs to tear-down the machine after processing are zero. Mason and Anderson [90] show that problems with nonzero tear-downs can easily be converted into problems with sequence dependent setups and zero tear-downs; hence the latter assumption is not restrictive. Together with the triangle inequality, zero tear-downs mean that a setup from the idle machine is larger than from any other family, i.e. $st_{0,i} \geq st_{g,i}, \forall g, i = 1, \ldots, N$, analogously for setup costs. The assumption about the time distance between deadlines means, that the decision of whether or not to batch two jobs is not a trivial one. E.g. if $d_{(i,j)} = d_{(i,j+1)}$, we would concatenate $(i,j)$ and $(i, j+1)$ to one job. This assumption may be interpreted as the result of an aggregation step: a job $(i,j)$

---

[4]In the case of unit time jobs, i.e. $p_{(i,j)} \equiv 1$, this assumption is relaxed to $d_{(i,j)} + p_{(i,j+1)} \leq d_{(i,j+1)}$.

is specified such that we cannot decide a priori of whether to batch it with job $(i, j - 1)$ or $(i, j + 1)$, $1 < j < n_i$.

The following definitions are useful for a precise description of the models and algorithms.

**Definition 2.1** *A sequence* $\pi = ((i_{[1]}, j_{[1]}), (i_{[2]}, j_{[2]}), \ldots, (i_{[k]}, j_{[k]}), \ldots, (i_{[J]}, j_{[J]}))$ *assigns each job* $(i, j)$ *to a position* $k$ *and vice versa.*

Similar to an address operator in programming, we index the position $k$ in brackets.

**Definition 2.2** *A schedule* $\sigma = (C_{(i_{[1]}, j_{[1]})}, C_{(i_{[2]}, j_{[2]})}, \ldots, C_{(i_{[k]}, j_{[k]})}, \ldots, C_{(i_{[J]}, j_{[J]})})$ *is a vector of completion times.*

A sequence only denotes the processing order whereas a schedule also contains the times at which the jobs complete. A schedule $\sigma$ is called *feasible* (for a certain model) if $\sigma$ satisfies the constraints of the model; a feasible schedule is also called a *solution*. A solution $\sigma$ is *optimal* if no other solution with a smaller objective function value exists. The different models are presented in the next section (cf. Tables 2.4, 2.5 and 2.9).

Terms describing a schedule will be used as follows (maintaining the direction of the time axis): if a job $(i^1, j^1)$ is sequenced and scheduled *before* job $(i^2, j^2)$, then job $(i^1, j^1)$ is scheduled at an earlier position or completion time, i.e. $C_{(i_{[k]}^1, j_{[k]}^1)} < C_{(i_{[l]}^2, j_{[l]}^2)}$, and $k < l$. We will say job $(i^1, j^1)$ is consecutively *sequenced* before job $(i^2, j^2)$ if job $(i^2, j^2)$ is at the *next* position, i.e. $l = k+1$. If we consecutively *schedule* job $(i^1, j^1)$ before $(i^2, j^2)$, there is no idle time between both jobs, i.e. $C_{(i_{[k]}^1, j_{[k]}^1)} + st_{(i_{[k]}^1, i_{[k+1]}^2)} + p_{(i_{[k+1]}^2, j_{[k+1]}^2)} = C_{(i_{[k+1]}^2, j_{[k+1]}^2)}$. *Leftshifting* a job means to schedule a job earlier, in a *local leftshift* the sequence is maintained, in a *global leftshift* the sequence of jobs is changed (and rightshift is just the opposite). We will use "$\rightarrow$" to abbreviate "from... to...", e.g. $g \rightarrow i$ denotes "from (family) $g$ to (family) $i$".

The following definitions allow a concise definition of batches for the different batching types:

**Definition 2.3** *A partial sequence of consecutively sequenced jobs of one family is called a group.*

**Definition 2.4** *A partial schedule of consecutively scheduled jobs is called a block.*

In a group, all jobs belong to one family and they may be scheduled with inserted idle time. In a block, jobs may belong to different families, but a block does not contain idle time. Both group and block may consist of only one single job.

**Definition 2.5** *A batch is defined as follows:*

1. *item availability–preemptive batching (ia-pb): each group is a batch.*

2. *item availability–nonpreemptive batching (ia-npb) and batch availability (ba): all jobs in a group which form a block without any setup belong to one batch.*

With the following definition of occupation we derive a necessary condition for the feasibility of a schedule.

**Definition 2.6** *The schedule constructed by the following algorithm is called occupation.*

1. *Sort all jobs in order of increasing deadlines.*

2. *Divide all processing times by the number of identical machines.*[5]

3. *Schedule the jobs on one machine beginning with the last job and rightshift them as far as possible without considering setups.*

The occupation gives the smallest set of the time instances in which the machine(s) are "occupied", see e.g. Figures 2.3 or 2.5. A necessary condition for feasibility is that, in the occupation, the first job must not start earlier than zero, and that condition is also sufficient for zero setup times.

In order to relate to the variants of the BSP in a concise manner, we introduce the descriptor presented in Table 2.2. The descriptor has three fields $[\alpha/\beta/\gamma]$ as is common in machine scheduling[6]: $\alpha$ refers to the machine, $\beta$ to the jobs and $\gamma$ to the objective.

---

[5] Processing times may now no longer be integer valued. However, integrality of the processing times is not needed to perform the next step of the algorithm.

[6] For an introduction to the classical three-field descriptors for scheduling problems cf. e.g. Pinedo [99] or Domschke et al. [39].

Existing descriptors do not capture all BSP characteristics (cf. e.g. Pinedo [99], p. 15), therefore, some new definitions are needed in this context. We consider family scheduling instead of general setups, and we must distinguish between batching types. So we use the entry $fam$ and the batching type in the $\beta$-field. The $\alpha$ field refers to the machine characteristics. We distinguish between single ($\alpha = 1$) and identical parallel machines ($\alpha = P$). For multi-level product structures we consider either the single-machine ($\alpha = ML1$) or the multi-machine ($\alpha = ML$ or $F$) case. For $\alpha = ML$ we assume a fixed family-machine assignment (and therefore also each job is assigned to a machine), but there exist precedence constraints between jobs. $\alpha = F$ denotes the flow-shop case with identical machines.

The characteristics of jobs and families are represented in the $\beta$-tuple: the entry $\beta = fam$ denotes family scheduling, and the batching type is given in entry $\beta_1$. Entry $\beta_2$ denotes sequence dependent setups by $st_{g,i}$ and sequence independent setups by $st_i$. In $\beta_3$, jobs have a deadline $d_{(i,j)}$ and are ready at time zero. Otherwise, jobs have release dates $r_{(i,j)}$ (processing must start after $r_{(i,j)}$) and there is a common deadline $D$ for all jobs. The general case, where jobs have release dates $and$ deadlines is not considered. We may use an entry $\beta_4$ for other cases, which is explained when needed, e.g. $p_{(i,j)} = 1$ if all processing times equal one.

The scheduling objective is represented in the $\gamma$ field. As one objective for an optimization problem we consider the minimization of holding costs, which is equivalent to minimize weighted earliness. The earliness depends on the completion time $C_{(i,j)}$, therefore we will use $\gamma = \sum w_{(i,j)} C_{(i,j)}$ in the descriptor. The minimization of setup costs is denoted by $\gamma = \sum sc_{g,i}$, the sum of earliness and setup costs with $\gamma = *$. If the entry is omitted, we look at the (in general not trivial) feasibility problem.

In the following section several models are presented which illustrate the three-field descriptor.

## 2.3   Single-Machine Case

In this section we present model formulations for the single-machine case ($\alpha = 1$). The decision variables are shown in Table 2.3. $\pi$ is the job sequence, the job at position $k$

**Table 2.2:** *Three-Field [α/β/γ] Descriptor for the BSP*

| **Machine** | | | |
|---|---|---|---|
| $\alpha \in \{1, F, ML1, ML, P\}$ | 1 | : | single-machine |
| | $F$ | : | flow-shop with identical machines |
| | $ML1$ | : | multi-level, single-machine |
| | $ML$ | : | multi-level, multi-machine, fixed family-machine assignment |
| | $P$ | : | identical parallel machines, single-level |
| **Job** | | | |
| $\beta = (fam, \beta_1, \beta_2, \beta_3)$ | | | |
| $\beta_1 \in \{ia\text{-}pb, ia\text{-}npb, ba, *\}$ | $ia\text{-}pb$ | : | item availability-preemptive batching |
| | $ia\text{-}npb$ | : | item availability-nonpreemptive batching |
| | $ba$ | : | batch availability |
| | $*$ | : | $ia\text{-}pb$, $ia\text{-}npb$ or $ba$ |
| $\beta_2 \in \{st_{g,i}, st_i, *\}$ | $st_{g,i}$ | : | sequence dependent setup times |
| | $st_i$ | : | sequence independent setup times |
| | $*$ | : | $st_{g,i}$ or $st_i$ |
| $\beta_3 \in \{d_{(i,j)}, r_{(i,j)}D, *\}$ | $d_{(i,j)}$ | : | each job has an individual deadline |
| | $r_{(i,j)}D$ | : | each job has an individual release date and a common deadline is given for all jobs |
| | $*$ | : | $d_{(i,j)}$ or $r_{(i,j)}D$ |
| **Objective** | | | |
| $\gamma \in \{\sum w_{(i,j)}C_{(i,j)}, \sum sc_{g,i}, \sum sc_i, *, \}$ | $\sum w_{(i,j)}C_{(i,j)}$ | : | weighted completion time |
| | $\sum sc_{g,i}$ | : | sequence dependent setup costs |
| | $\sum sc_i$ | : | sequence independent setup costs |
| | $*$ | : | $\sum w_{(i,j)}C_{(i,j)} + \sum sc_{g,i}$ or $\sum w_{(i,j)}C_{(i,j)} + \sum sc_i$ |
| | | : | no objective specified, constraint satisfaction or feasibility problem |

**Table 2.3:** *Decision Variables of the BSP*

| | | |
|---|---|---|
| $\pi$ | : | Sequence of all jobs, |
| | | $\pi = ((i_{[1]}, j_{[1]}), (i_{[2]}, j_{[2]}), \ldots, (i_{[k]}, j_{[k]}), \ldots, (i_{[J]}, j_{[J]}))$ |
| $(i_{[k]}, j_{[k]})$ | : | denotes the job at position $k$ |
| $C_{(i,j)}$ | : | completion time of job $(i,j)$ |
| $P_k$ | : | $\begin{cases} 1 & \text{if idle time preempts production between jobs} \\ & (i_{[k-1]}, j_{[k-1]}) \text{ and } (i_{[k]}, j_{[k]}), \ k = 2, \ldots, J \\ 0 & \text{otherwise} \end{cases}$ |

is denoted by $(i_{[k]}, j_{[k]})$, and the completion time by $C_{(i_{[k]}, j_{[k]})}$. Using a tuple to index the jobs we are also able to access the family $i_{[k]}$ scheduled at position $k$. Table 2.4 presents a model for *ia-pb*. In this "conceptual" model, the index is a decision variable; therefore, the model cannot be solved with conventional mixed–integer programming (MIP) solvers. However, this formulation of the model will be helpful for the presentation of algorithms. Other formulations, which can be solved by MIP solvers, are presented in Chapter 6.

The feasibility problem is expressed by constraints (2.2) to (2.4). In constraints (2.2) the first element in the "min"-term models the materials flow transformation constraints: jobs sequenced at positions $k-1$ and $k$ must not overlap and the family setup must be taken into account (disjunction). The second element in the minimization models the service constraints: each job must be available before its deadline, which means that a job must complete before its deadline for *ia-pb*. Constraints (2.3) and (2.4) initialize the beginning and the end of the schedule, respectively.

The scheduling objective is given in (2.1). The second term counts the setup costs incurred if families are switched from $i_{[k-1]}$ to $i_{[k]}$ between consecutively sequenced jobs. In the first term of (2.1), for each job the early completion (or the flow time between completion time and deadline) is weighted. As deadlines are met, the earliness is always nonnegative. Furthermore, $w_{(i,j)} d_{(i,j)}$ is a constant. Therefore we *maximize* weighted completion times in (2.1). For an easier comparison with lotsizing models in Chapter 4 we keep (2.1) as the objective function, and $\gamma = \sum w_{(i,j)} C_{(i,j)}$ denotes the minimization of weighted earliness.

**Table 2.4:** *Model for* $[1/fam, ia\text{-}pb, st_{g,i}, d_{(i,j)} / \sum w_{(i,j)} C_{(i,j)} + \sum sc_{g,i}]$

$$\text{Min } Z_{BSP} = \sum_{k=1}^{J} w_{(i_{[k]}, j_{[k]})} \left( d_{(i_{[k]}, j_{[k]})} - C_{(i_{[k]}, j_{[k]})} \right) + sc_{i_{[k-1]}, i_{[k]}} \tag{2.1}$$

subject to

$$C_{(i_{[k-1]}, j_{[k-1]})} \leq \min \left\{ C_{(i_{[k]}, j_{[k]})} - p_{(i_{[k]}, j_{[k]})} - st_{i_{[k-1]}, i_{[k]}}, d_{(i_{[k-1]}, j_{[k-1]})} \right\} \quad k = 1, \ldots, J \tag{2.2}$$

$$(i_{[0]}, j_{[0]}) = (0,0); \quad d_{(0,0)} = C_{(0,0)} = 0 \tag{2.3}$$

$$C_{(i_{[J]}, j_{[J]})} = d_{(i_{[J]}, j_{[J]})} \tag{2.4}$$

In Table 2.5 a model for *ia-npb* is presented. $B$ denotes a big number. The decision variable $P_k$ determines whether consecutively sequenced jobs are also consecutively scheduled or not. $P_k$ is determined within the constraints (2.7): if there is idle time between job $(i_{[k-1]}, j_{[k-1]})$ and $(i_{[k]}, j_{[k]})$, the term in brackets is positive and $P_k$ is set to one. Then there is a setup $0 \to i_{[k]}$ from the idle machine instead of a setup $i_{[k-1]} \to i_{[k]}$ in constraints (2.6) and in the objective (2.5). Note that, due to Remark 2.1, the terms $(st_{0,i} - st_{g,i})$ and $(sc_{0,i} - sc_{g,i})$ are always nonnegative.

A model for *ba* requires technical overhead without providing new insights and is thus not presented here.

In the following we illustrate the above models with a numerical example. Table 2.6 and Table 2.7 display the parameters: setup times $st_{g,i}$, setup costs $sc_{g,i}$ and holding costs $h_i$ for the $N = 3$ families are given in Table 2.6. We have $sc_{g,i} = 5st_{g,i}$. There are $n_i = 2, 2, 3$ jobs, and the job attributes deadlines $d_{(i,j)}$, processing times $p_{(i,j)}$ and weights $w_{(i,j)}$ are given in Table 2.7. Figure 2.3 depicts the jobs of the 3 families scheduled at their deadlines, i.e. for each job $p_{(i,j)}$ and $d_{(i,j)}$ are displayed. The occupation of this instance is given on top of Figure 2.3

For *ia-pb* the sequence and schedule $\sigma_I$ in Table 2.7 represent the optimal solution. The (consecutively sequenced) jobs of family 3 are one group and one batch with idle time in

**Table 2.5:** *Model for [1/fam, ia-npb,$st_{g,i}$,$d_{(i,j)}$/ $\sum w_{(i,j)}C_{(i,j)}$+ $\sum sc_{g,i}$]*

---

$$\text{Min } Z_{BSP} = \sum_{k=1}^{J} w_{(i_{[k]},j_{[k]})}(d_{(i_{[k]},j_{[k]})} - C_{(i_{[k]},j_{[k]})}) + sc_{i_{[k-1]},i_{[k]}} + P_k(sc_{0,i_{[k]}} - sc_{i_{[k-1]},i_{[k]}}) \quad (2.5)$$

subject to

$$C_{(i_{[k-1]},j_{[k-1]})} \leq \min\left\{C_{(i_{[k]},j_{[k]})} - p_{(i_{[k]},j_{[k]})} - st_{i_{[k-1]},i_{[k]}} - P_k(st_{0,i_{[k]}} - st_{i_{[k-1]},i_{[k]}}), d_{(i_{[k-1]},j_{[k-1]})}\right\}$$
$$k = 1,\dots,J \qquad (2.6)$$

$$BP_k - (C_{(i_{[k]},j_{[k]})} - p_{(i_{[k]},j_{[k]})} - st_{i_{[k-1]},i_{[k]}} - C_{(i_{[k-1]},j_{[k-1]})}) \geq 0 \qquad k = 1,\dots,J \qquad (2.7)$$

$$P_k \in \{0,1\} \qquad k = 1,\dots,J \qquad (2.8)$$

$$(i_{[0]}, j_{[0]}) = (0,0); \quad d_{(0,0)} = C_{(0,0)} = 0 \qquad (2.9)$$

$$C_{(i_{[J]},j_{[J]})} = d_{(i_{[J]},j_{[J]})} \qquad (2.10)$$

---

**Table 2.6:** *Numerical Example: Setup and Holding Costs*

| $st_{g,i}$ | 1 | 2 | 3 | $sc_{g,i}$ | 1 | 2 | 3 | $h_i$ |
|---|---|---|---|---|---|---|---|---|
| 0 | 1 | 2 | 1 | 0 | 5 | 10 | 5 |  |
| 1 | 0 | 1 | 1 | 1 | 0 | 5 | 5 | 1 |
| 2 | 0 | 0 | 1 | 2 | 0 | 0 | 5 | 1 |
| 3 | 1 | 2 | 0 | 3 | 5 | 10 | 0 | 1 |

**Table 2.7:** *Numerical Example: Job Attributes and Schedule $\sigma_I$*

| $(i,j)$ | $(1,1)$ | $(1,2)$ | $(2,1)$ | $(2,2)$ | $(3,1)$ | $(3,2)$ | $(3,3)$ |
|---|---|---|---|---|---|---|---|
| $d_{(i,j)}$ | 8 | 21 | 9 | 20 | 10 | 16 | 21 |
| $p_{(i,j)}$ | 1 | 2 | 1 | 1 | 1 | 2 | 1 |
| $w_{(i,j)}$ | 1 | 2 | 1 | 1 | 1 | 2 | 1 |
| $k$ | 1 | 2 | 3 | 4 | 5 | 6 | 7 |
| $(i_{[k]}, j_{[k]})$ | $(2,1)$ | $(1,1)$ | $(3,1)$ | $(3,2)$ | $(3,3)$ | $(2,2)$ | $(1,2)$ |
| $C_{(i_{[k]}, j_{[k]})}$ | 7 | 8 | 10 | 15 | 16 | 19 | 21 |

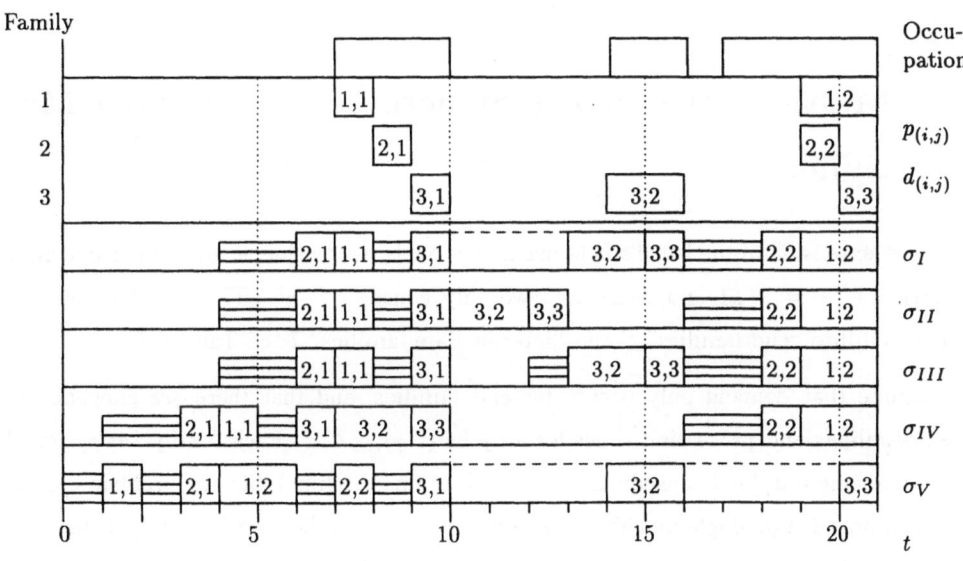

**Figure 2.3:** *Example for [1/fam ,*,$st_{g,i}$,$d_{(i,j)}$/*]*

the batch. The setup costs are determined by the sequence of families: $0 \to 2 \to 1 \to 3 \to 2 \to 1$, and the earliness of each job depends on $C_{(i,j)}$. E.g., the earliness of job $(2,1)$ is $9 - 7 = 2[TU]$, the optimal objective function value of $\sigma_I$ is $35[MU]$.

For *ia-npb* both $\sigma_{II}$ and $\sigma_{III}$ represent a solution, jobs in a batch must be consecutively *scheduled*. Setup costs in $\sigma_{II}$ equal the setup costs of $\sigma_I$, but there are higher earliness costs because $(3,2)$ and $(3,3)$ are leftshifted to be batched with $(3,1)$. In $\sigma_{III}$ earliness costs equal those of $\sigma_I$, but setup costs are higher because the group $(3,1)$, $(3,2)$ and $(3,3)$ forms two batches. For the cost parameters in Table 2.6, $\sigma_{III}$ with an objective function value of $40[MU]$ is optimal (versus $44[MU]$ for $\sigma_{II}$).

For *ba* both $\sigma_{III}$ and $\sigma_{IV}$ represent solutions. A job becomes available after batch completion. Thus, if all jobs of family 3 form a batch (as in $\sigma_{IV}$), the batch must complete at time $d_{(3,1)} = 10$ so that $(3,1)$ is available at its deadline. In *ba* all jobs can be rightshifted so that there is no idle time in a batch, because the last job in a batch must complete before deadline of the first job in the batch.

## 2.4   Multi-Level Product Structures and Parallel Machines

As a first extension of the basic single-machine models we consider **multi-level product structures** ($\alpha = ML1$ and $ML$), and we only consider *ia-pb*. The set of families is partitioned into "end families" $\mathcal{E}$ and "intermediate families" $\mathcal{I}$, cf. Table 2.8.

We assume that demand only occurs for end families, and that there are exogenously given deadlines and processing times for all jobs $(i,j)$, $i \in \mathcal{E}$. Each family $i$ is assigned to one machine $ma_i$ (and consequently, each job to one machine), but the problem cannot be decomposed into single-machine problems, because families are linked by a product structure, which is represented by a gozinto graph. The gozinto graph may be *linear* (each family has at most one successor and one predecessor), *convergent* (each family has at most one successor) or *general* (a family may have more than one successor). Each general gozinto graph can be decomposed into a convergent one. This is illustrated in the

**Table 2.8:** *BSP Parameters for Multi-Level Product Structures*

| | | |
|---|---|---|
| $\mathcal{E}$ | : | set of end families |
| $\mathcal{I}$ | : | set of intermediate families, $\mathcal{E} \cup \mathcal{I} = \{1, \ldots, N\}$ |
| $u_i$ | : | unique successor of family $i$, $i \in \mathcal{I}$ |
| $a_{i,u_i}$ | : | gozinto factor, units of $i$ needed for one unit of $u_i$ |
| $m$ | : | machine index $m = 1, \ldots, M$ |
| $ma_i$ | : | machine $m$ on which family $i$ has to be produced |

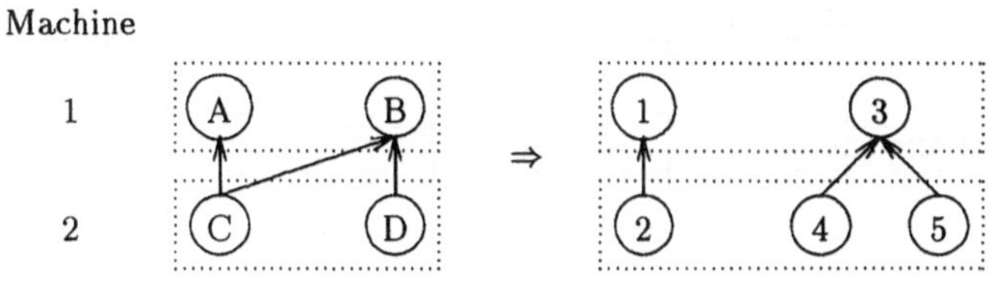

**Figure 2.4:** *Gozinto Graph and Machine Assignment in Figure 2.5*

following example. A formal description of the decomposition (which is done recursively) is given in Afentakis and Gavish [1] or Koppelmann [76].

In multi-level models, demand of end families results in (internal) demand of intermediate families. For dealing with a gozinto graph within the BSP, the basic idea is as follows: a job $(i,j)$ $i \in \mathcal{I}$ is derived from $(u_i, j)$; jobs $(i,j)$ $i \in \mathcal{E}$ are exogenously given, and the attributes $p_{(i,j)}$ and $d_{(i,j)}$ of jobs $(i,j)$ $i \in \mathcal{I}$ are derived recursively. The gozinto graph is convergent; hence, for each $(i,j)$ there is a unique successor job $(u_i, j)$. Then, the gozinto graph is represented by precedence constraints between jobs, which must additionally be taken into account in scheduling.

An example for a multi-level product structure is depicted in Figure 2.4. There are end families $\mathcal{E} = \{A, B\}$ and intermediate families $\mathcal{I} = \{C, D\}$. The gozinto factor $a_{i,u_i}$ is equal to one for all arcs except of arc $C \rightarrow B$ for which it is 0.5. This general product structure is now decomposed into a convergent product structure as follows: for family $C$,

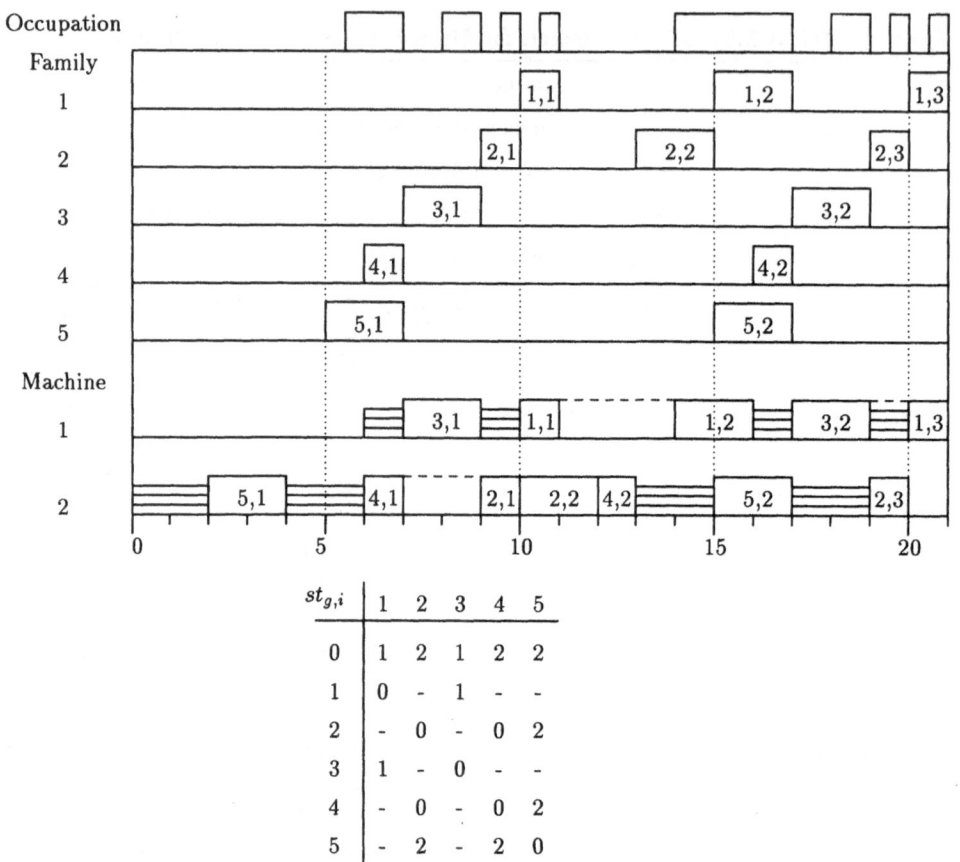

**Figure 2.5:** *Example for [ML/fam,ia-pb,$st_{g,i}$,$d_{(i,j)}$/*]*

there are two *paths* to the two end families $A$ and $B$, thus $C$ is decomposed into two
(artificial) families $2, 4$. Figure 2.4 also shows the family–machine assignment $ma_i$, e.g.
the two (artificial) families $2, 4$ are assigned to machine 2. Figure 2.5 displays the jobs
at their deadlines, a solution in terms of a Gantt-chart and the matrix of setup times.
Setups occur only between families on the same machine and are otherwise denoted with
"$-$" in $st_{g,i}$, e.g. there is no setup $2 \to 1$ because $ma_1 = 1 \neq 2 = ma_2$. Furthermore,
between the (artificial) families $2, 4$, we have $st_{2,4} = st_{4,2} = 0$. In the convergent product

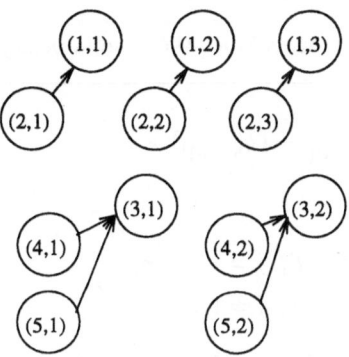

**Figure 2.6:** *One Successor for each Job in $[ML/fam,ia\text{-}pb,st_i,d_{(i,j)}/*]$*

structure, each intermediate family $i \in \mathcal{I}$ has one unique successor family $u_i$. Given the $n_1 = 3, n_3 = 2$ jobs for the end families 1 and 3, we derive $n_2 = 3$, $n_4 = n_5 = 2$ jobs for the intermediate families 2, 4 and 5 with the gozinto factor $a_{i,u_i}$. E.g., for family $3 \in \mathcal{E}$, given $p_{(3,1)} = 2$ we derive $p_{(4,1)} = p_{(3,1)} \cdot a_{4,u_4} = 2 \cdot 0.5 = 1$ and $d_{(4,1)} = d_{(3,1)} - p_{(3,1)} = 7$. The occupation for the instance is given at the top of Figure 2.5.[7] The convergent product structure translates into precedence constraints between each job $(i,j)$ and its successor job $(u_i, j)$. Thus, we obtain a BSP with additional precedence constraints between jobs. Cf. Figure 2.6, family 2 "goes into" family 1, and hence, there is an arc from $(2,1)$ to $(1,1)$.

The time distances associated with the arcs explicitly model the transfer between machines. We assume a value of zero to facilitate analysis, however, transportation times for instance (which may differ among families) could easily be integrated into the model in this way.

Table 2.9 (with the parameters of Table 2.8) represents a model for the multi-machine multi-level problem with a convergent product structure. Constraints (2.12) to (2.15) model the feasibility problem $[ML/fam,st_i,d_{(i,j)},ia\text{-}pb/]$. In constraints (2.12) we derive the "multi-level deadline" $\tilde{d}_{(i,j)}$, which is $d_{(i,j)}$ for end families $\mathcal{E}$, and the start time $C_{(u_i,j)} - p_{(u_i,j)}$ of the successor job $(u_i, j)$ for intermediate families $\mathcal{I}$. The decision variables in Table 2.9 are the same as for the single-machine models: we decide about one

---

[7]Note that processing times must be divided by $M = 2$.

**Table 2.9:** *Model for* $[ML/fam,ia\text{-}pb,st_{g,i},d_{(i,j)}/*]$

---

$$\text{Min } Z_{BSP} = \sum_{k=1}^{J} w_{(i_{[k]},j_{[k]})}(d_{(i_{[k]},j_{[k]})} - C_{(i_{[k]},j_{[k]})}) + sc_{i_{[z(k)]},i_{[k]}} \tag{2.11}$$

subject to

$$\tilde{d}_{(i,j)} = \begin{cases} d_{(i,j)} & , \quad i \in \mathcal{E}, \ j = 1,\ldots,n_i \\ C_{(u_i,j)} - p_{(u_i,j)} & , \quad i \in \mathcal{I}, \ j = 1,\ldots,n_i \end{cases} \tag{2.12}$$

$$C_{(i_{[z(k)]},j_{[z(k)]})} \le \min\left\{C_{(i_{[k]},j_{[k]})} - p_{(i_{[k]},j_{[k]})} - st_{i_{[z(k)]},i_{[k]}}, \tilde{d}_{(i_{[z(k)]},j_{[z(k)]})}\right\} \quad k = 1,\ldots,J \tag{2.13}$$

$$(i_{[0]},j_{[0]}) = (0,0); \ d_{(0,0)} = C_{(0,0)} = 0 \tag{2.14}$$

$$C_{(i_{[J]},j_{[J]})} = d_{(i_{[J]},j_{[J]})} \tag{2.15}$$

---

overall sequence $\pi$ and the completion times $C_{(i,j)}$. The fixed family–machine assignment $ma_i$ determines for each job the machine on which it must be scheduled. But here, the sequencing constraints apply separately for each machine, and we must derive the job sequence on each machine from the overall sequence $\pi$. To do so, we need an auxiliary index variable $z(k)$ which is defined as follows:

$$z(k) = \begin{cases} \text{position in } \pi \text{ of the job preceding } (i_{[k]},j_{[k]}) \text{ on the same machine} \\ \text{position 0 if } (i_{[k]},j_{[k]}) \text{ is the start job on the machine} \end{cases}$$

Thus, jobs $(i_{[z(k)]},j_{[z(k)]})$ and $(i_{[k]},j_{[k]})$ are sequenced consecutively on the same machine. The sequencing constraints (2.13) are now expressed between $[z(k)]$ and $[k]$ and not between $[k-1]$ and $[k]$ as in the single-machine case. Each job appears two times in constraints (2.13), the first time as $(i_{[z(k)]},j_{[z(k)]})$ and the second time as $(i_{[k]},j_{[k]})$.

The objective (2.11) calculates *echelon stock* costs (cf. Domschke et al. [39]): also for jobs $(i,j)$ $i \in \mathcal{I}$ the earliness costs are calculated as $w_{(i,j)}(d_{(i,j)}-C_{(i,j)})$. Consider Figure 2.7 as an example. $(2,1)$ goes into $(1,1)$, and $p_{(2,1)}$, $d_{(2,1)}$ are derived. Let the completion times as in the bottom line of the figure, and $h_i = 1$, then $p_{(i,j)} = w_{(i,j)}$. Inventory of job

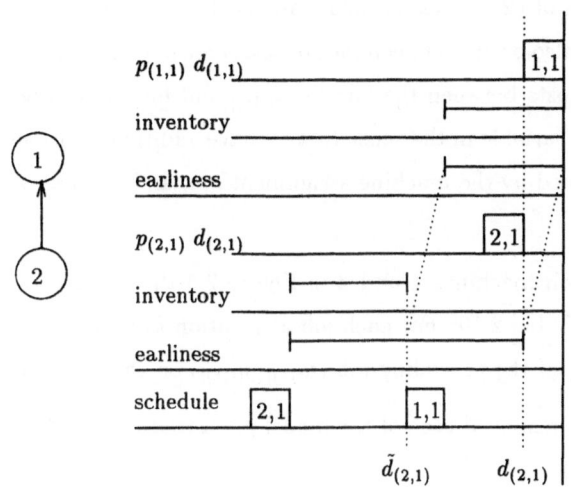

**Figure 2.7:** *Objective (2.11) calculates costs of the echelon stock*

**Table 2.10:** *Solution of the [ML/fam,ia-pb,$st_{g,i}$,$d_{(i,j)}$/*] Example of Figure 2.5*

| $[k]$ | 1 | 2 | 3 | 4 | 5 | 6 |
|---|---|---|---|---|---|---|
| $(i_{[k]}, j_{[k]})$ | $(5,1)$ | $(4,1)$ | $(3,1)$ | $(2,1)$ | $(1,1)$ | $(2,2)$ |
| $C_{(i_{[k]}, j_{[k]})}$ | 4 | 7 | 9 | 10 | 11 | 12 |
| $z(k)$ | 0 | 1 | 0 | 2 | 3 | 4 |
| $(i_{[z(k)]}, j_{[z(k)]})$ | $(0,0)$ | $(5,1)$ | $(0,0)$ | $(4,1)$ | $(3,1)$ | $(2,1)$ |
| $[k]$ | 7 | 8 | 9 | 10 | 11 | 12 |
| $(i_{[k]}, j_{[k]})$ | $(4,2)$ | $(1,2)$ | $(5,2)$ | $(3,2)$ | $(2,3)$ | $(1,3)$ |
| $C_{(i_{[k]}, j_{[k]})}$ | 13 | 16 | 17 | 19 | 20 | 21 |
| $z(k)$ | 6 | 5 | 7 | 8 | 9 | 10 |
| $(i_{[z(k)]}, j_{[z(k)]})$ | $(2,2)$ | $(1,1)$ | $(4,2)$ | $(1,2)$ | $(5,2)$ | $(3,2)$ |

$(2,1)$ occurs between $C_{(2,1)}$ and $\tilde{d}_{(2,1)}$. But the echelon stock of $(2,1)$ is between $C_{(2,1)}$ and $d_{(2,1)}$ (= the earliness of $(2,1)$), $(2,1)$ incurs an echelon stock even after $(2,1)$ "has gone into" $(1,1)$.[8] Setups also occur between jobs on the same machine, and the objective $(2.11)$ considers the setup costs between the families $i_{[z(k)]}$ and $i_{[k]}$. It is important to note that $z(k)$ is not a decision variable in the sense that we have additional options in scheduling the jobs; $z(k)$ is determined by the machine assignment $ma_i$ and $\pi$, so a solution is described by $\pi$ only.

For the Gantt-chart on machines 1 and 2 in Figure 2.5 the solution in terms of decision variables is given in Table 2.10. For each job at position $k$, its completion time $C_{(i_{[k]}, j_{[k]})}$, and the position $z(k)$ of the preceding job $(i_{[z(k)]}, j_{[z(k)]})$ on the same machine are given.

The flow-shop case is a special case of the model presented here and is examined in more detail in Section 3.4.4.

We model $[ML/fam,ia\text{-}pb,st_{g,i},d_{(i,j)},/*]$ with one overall sequence instead of $M$ single-machine sequences. This allows us to formulate a structural property in Section 2.6, which is useful in the algorithmic analysis in Chapter 5.

As a second extension of the single-machine models we consider identical **parallel machines** ($\alpha = P$) with single-level production. Regardless of the machine on which a specific job is scheduled, the processing time is always the same.[9]

In addition to the sequencing decision, we now have to make an *allocation* decision on which machine to schedule a job. Machines are indexed by $m = 1, \ldots, M$, and the

---

[8]Echelon stock costs $W_{ech}$ in Figure 2.7 are

$$W_{ech} = w_{(2,1)}(d_{(2,1)} - C_{(2,1)}) + w_{(1,1)}(d_{(1,1)} - C_{(1,1)})$$

We have $d_{(2,1)} = d_{(1,1)} - p_{(1,1)}$ and $\tilde{d}_{(2,1)} = C_{(1,1)} - p_{(1,1)}$. However, we can express $W_{ech}$ in terms of (physical) inventories as well, i.e.

$$W_{ech} = w_{(2,1)}(\tilde{d}_{(2,1)} - C_{(2,1)}) + (w_{(1,1)} + w_{(2,1)})(d_{(1,1)} - C_{(1,1)}).$$

Now the earliness of $(1,1)$ is weighted with $(w_{(1,1)} + w_{(2,1)})$. Equivalently, in geometric terms: the earliness minus inventory of $(2,1)$ is the earliness of $(1,1)$, see the dotted lines in Figure 2.7.

[9]For the more general cases with *uniform* and *heterogeneous* machines, cf. e.g. Domschke et al. [39].

**Table 2.11:** *Values of the Decision Variables of the (Feasible) Schedule $\sigma$ of Figure 2.8*

| $[k]$ | 1 | 2 | 3 | 4 | 5 | 6 | 7 | 8 | 9 |
|---|---|---|---|---|---|---|---|---|---|
| $(i_{[k]}, j_{[k]})$ | $(1,1)$ | $(4,1)$ | $(2,1)$ | $(5,1)$ | $(3,1)$ | $(2,2)$ | $(1,2)$ | $(4,2)$ | $(1,3)$ |
| $MA_{[k]}$ | 1 | 2 | 1 | 2 | 1 | 2 | 1 | 2 | 1 |
| $C_{(i_{[k]}, j_{[k]})}$ | 4 | 4 | 5 | 6 | 7 | 10 | 12 | 13 | 13 |
| $z(k)$ | 0 | 0 | 1 | 2 | 3 | 4 | 5 | 6 | 7 |
| $(i_{[z(k)]}, j_{[z(k)]})$ | $(0,0)$ | $(0,0)$ | $(1,1)$ | $(4,1)$ | $(2,1)$ | $(5,1)$ | $(3,1)$ | $(2,2)$ | $(1,2)$ |
| $[k]$ | 10 | 11 | 12 | 13 | 14 | 15 | 16 | 17 | 18 |
| $(i_{[k]}, j_{[k]})$ | $(2,3)$ | $(4,3)$ | $(3,2)$ | $(5,2)$ | $(3,3)$ | $(4,4)$ | $(1,4)$ | $(2,4)$ | $(5,3)$ |
| $MA_{[k]}$ | 1 | 2 | 1 | 2 | 1 | 1 | 2 | 2 | 1 |
| $C_{(i_{[k]}, j_{[k]})}$ | 15 | 15 | 18 | 18 | 19 | 22 | 23 | 25 | 25 |
| $z(k)$ | 9 | 8 | 10 | 11 | 12 | 14 | 13 | 16 | 15 |
| $(i_{[z(k)]}, j_{[z(k)]})$ | $(1,3)$ | $(4,2)$ | $(2,3)$ | $(4,3)$ | $(3,2)$ | $(3,3)$ | $(5,2)$ | $(1,4)$ | $(4,4)$ |

additional decision variable $MA_k$[10] denotes the machine $m$ on which the job at position $k$ is scheduled. There are several special cases for the parallel machine problem which allow for an easier solution approach: First, if the number of machines $M$ no less than the number of families $N$ then each family can be dedicated to one machine such that the problem becomes trivial. We thus assume $M < N$ in the following. Second, if the workload of some families is so large that some machines can be assigned to these families for the whole planning horizon then we can reduce the problem by a preprocessing: we first schedule the jobs where one family is assigned to a machine for the whole planning horizon, and then consider a "remaining" problem with a smaller number of jobs and machines.

An exemplary instance for parallel machines is given in Figure 2.8. We depict the jobs of the $N = 5$ families at their deadlines and display a feasible schedule on the $M = 2$ machines with setups according to the setup time matrix. This schedule – in terms of decision

---

[10] $MA_k$ is an (uppercase) *decision variable* in contrast to the (lowercase) *parameter $ma_i$*.

Data: $p_{(i,j)}$, $d_{(i,j)}$

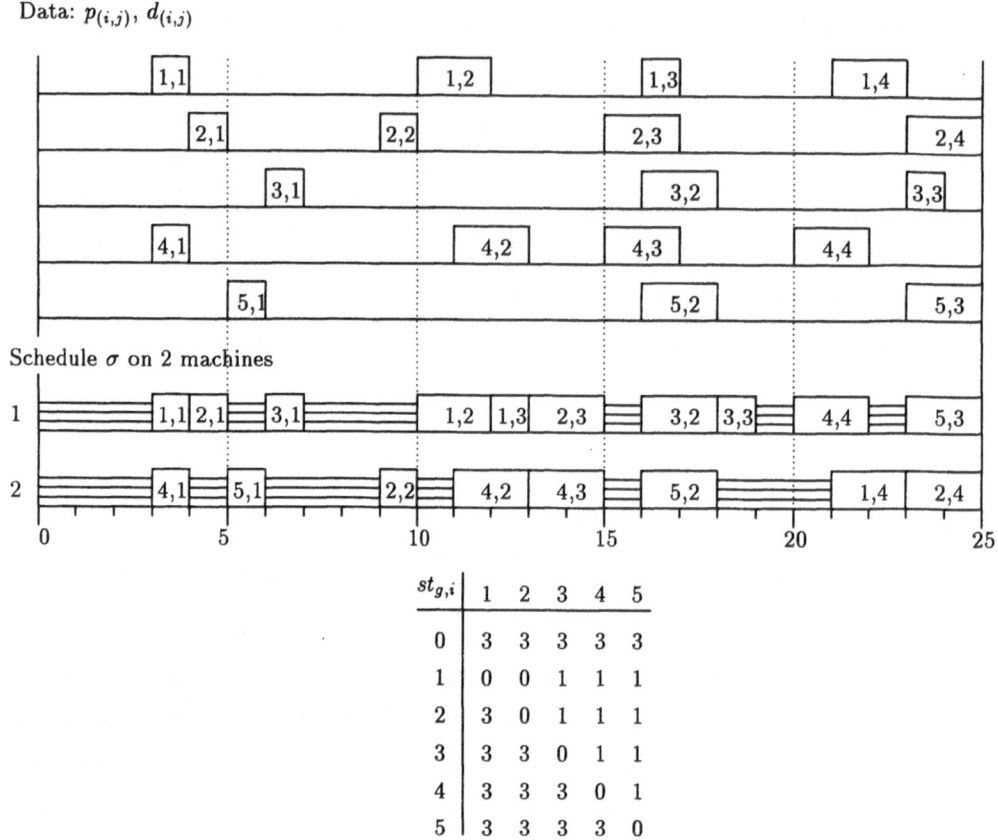

**Figure 2.8:** *Example for $[P/fam,ia\text{-}pb,st_i,d_{(i,j)}/]$*

variables – is shown in Table 2.11. Again we use an overall sequence $\pi$, but now the decision variable $MA_k$ also assigns the job at position $k$ to one of the $M = 2$ machines. The precedence operator $z(k)$ is now determined by $\pi$ and $MA_k$, while $z(k)$ again identifies the jobs $(i_{[z(k)]}, j_{[z(k)]})$ and $(i_{[k]}, j_{[k]})$ which are consecutively sequenced on the same machine.

We can model the parallel machine case in the same way as the $[1/fam,ia\text{-}pb,st_g,i,d_{(i,j)}/*]$ model in Table 2.4, simply by replacing $[k-1]$ with $[z(k)]$. The use of one overall sequence $\pi$ and $MA_k$ allows for a simple description of the solution space: all solutions are given by all *overall* sequences $\pi$ and machine assignments $MA_k$. This way, we can easily adapt an algorithm pertaining to the single-machine case to the case of identical parallel machines.

## 2.5 Complexity Analysis

In examining the relationship between the batching problems we use the three-field descriptor of Table 2.2. Figure 2.9 summarizes the complexity hierarchy of the $[\alpha/\beta/\gamma]$ descriptor. An arc between two entries means *"is a special case of"*. For instance, all problems with $\alpha = 1$ are a special case of the general case $\alpha = P$, and problems with $\alpha = P$ are at least as difficult to solve as those with $\alpha = 1$. For the complexity analysis, we can state that the general case is NP-hard if this is true for the special case; in this sense we may say that complexity results extend in the direction of the arc.

For the $\alpha$ entry it is easy to see that the single-machine case is also a special case of $F$ and $ML1$. For linear product structures with an identical machine assignment, $F$ is a special case of $ML$.

For the $\beta$ entry, problems with $\beta_2 = st_{g,i}$ include problems with $\beta_2 = st_i$ as a special case. For the hierarchy between the batching types, we know that the set of *ia-npb* schedules is a subset of the set of *ia-pb* schedules, and that the set of *ba* schedules is a subset of the set of *ia-npb* schedules; none of the sets is polynomially bounded. If there is no solution for problems with $\beta_1 = $ *ia-pb*, there will be no solution for $\beta_1 = $ *ia-npb* or for *ba*, either.[11] Thus, problems with $\beta_1 = $ *ia-npb* (*ba*) are at least as difficult to solve as for $\beta = $ *ia-pb* (*ia-npb*).

For the $\gamma$ entry, we know that results for the feasibility problem extend to the optimization problem, i.e. either to the minimization of weighted earliness ($\sum w_{(i,j)} C_{(i,j)}$) or to the minimization of setup costs. Clearly, if both objectives are combined in $*$, complexity results for each of the single objectives extend to the combined case.

Polynomial algorithms are known for some simple problems. In our analysis we need the results for the single-machine problem with release dates where the makespan must be minimized ($[1/r_j/C_{max}]$), and the single-machine problem where the sum of weighted completion times must be minimized ($[1//\sum w_j C_j]$). In $[1/r_j/C_{max}]$ and $[1//\sum w_j C_j]$

---

[11]Cf. Figure 2.3 for an illustration. $\sigma_{II}$ is feasible for *ia-npb* but not for *ba*.

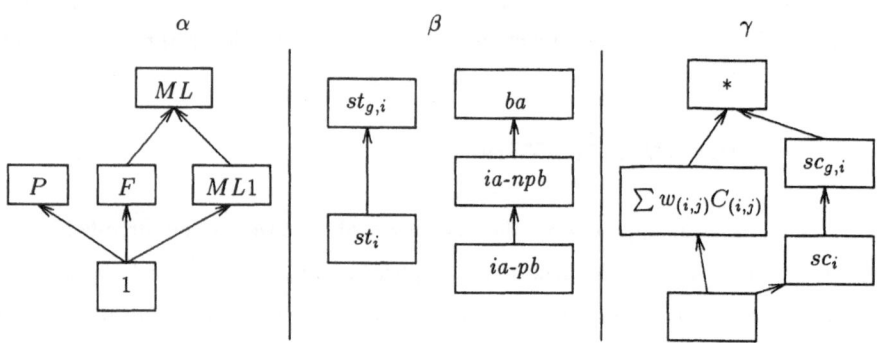

**Figure 2.9:** *The Complexity Hierarchy of the [α/β/γ] Descriptor*

there are no (family) setups,[12] jobs are independent and release dates $r_j$, weights $w_j$ and completion times $C_j$ are indexed with the subscript $j$. $[1/r_j/C_{max}]$ is solved ordering the jobs according to ERD (earliest release date). $[1//\sum w_j C_j]$ is solved with the Smith-Rule: order all jobs in SWPT (shortest weighted processing time) order.[13]

Unfortunately, the batch sequencing problem does not belong to the polynomially solvable cases. The complexity is best summarized in the following theorem.

**Theorem 2.1** *[1/fam,ia-pb,$st_i$ = 1,$d_{(i,j)}$/] is NP-complete for an arbitrary number N of families.*

**Proof:** Refer to Bruno and Downey [22] or Garey and Johnson [52] (p. 238), NP-completeness proof for problem SS6: SEQUENCING WITH DEADLINES AND SETUP TIMES.                                                                    □

Bruno and Downey prove NP-completeness for the above problem with sequence independent unit setup times ($st_i$ = 1) and 3 distinct deadlines reducing it to a knapsack problem. Thus already the "bottom most problem" of the complexity hierarchy is NP-complete.

Furthermore, NP-completeness of the feasibility implies NP-hardness of the corresponding optimization problem. For a related optimization problem without setups we have the

---

[12]We have no entry *fam* in $\beta$; consequently, the descriptor of Table 2.2 does not cover this case.

[13]Cf. e.g. Domschke et al. [39].

following result: the weighted completion time problem with individual processing times and deadlines (problem SS4 in Garey and Johnson [52]) is NP-complete in the strong sense, which holds also for proportional weights, cf. Swarcz and Liu [117] and Arkin and Roundy [7]. If we consider sequence dependent setups, even degenerate cases (e.g. a problem with one common deadline and one job per family) reduce to the (NP-hard) TSP for an arbitrary number of families. The results are summarized in the following corollary and can be proven using the complexity hierarchy in Figure 2.9.

**Corollary 2.1** *The problems $[\alpha/fam,*,*,d_{(i,j)}/*]$ are NP-hard for an arbitrary number $N$ of families for $\alpha \in \{1, F, ML1, ML, P\}$.*

So, the general statement is that the BSP is NP-hard. Consequently, efforts are warranted to derive structural properties to restrict the amount of search in algorithms, which is done in the following section.

## 2.6 Structural Properties

The analysis of the structural properties requires a clear distinction between a sequence $\pi$ and a schedule $\sigma$. Recall that in batching problems grouping *and* sequencing decisions are made simultaneously. A sequence $\pi$ does not describe a solution for all batching types which forces us to distinguish $\pi$ and $\sigma$.[14]

We generally observe that instead of considering the BSP with individual deadlines and zero release dates we can also consider the corresponding mirror problem. We mirror the time axis at the largest deadline, then deadlines convert into release dates $r_{(i,j)}$, and time zero becomes a common deadline $D$ for all jobs. Thus, for each $[\alpha/fam,*,d_{(i,j)},*/*]$ problem there is an equivalent $[\alpha/fam,*,r_{(i,j)}D,*/*]$ problem[15] for $\alpha \in \{1, F, ML1, ML, P\}$. We only need to change the signs of the weights and to transpose the setup matrices. As our main concern (according to Chapter 1) is to *deliver* jobs in time and to facilitate

---

[14]Cf. also French [51] (Section 2.2) for the distinction between sequence and schedule.

[15]Cf. also Bruno and Downey [22].

comparison with lotsizing models in Chapter 4, we keep the description with deadlines in the sequel. Therefore, we have to redefine the term *(semi)active.*

**Definition 2.7** *A schedule $\sigma$ for a sequence $\pi$ is called semiactive (active) if no job can be locally (globally) rightshifted maintaining (altering) the sequence $\pi$ and without leftshifting any other job.*

Definition 2.7 contrasts to the common definition without setups where jobs are *left*shifted (as for makespan problems).

Each sequence $\pi$ uniquely defines a semiactive schedule. The construction is easy using the constraints in the model for *ia-pb* in Table 2.4: starting with position $J$ (where $C_{(i,j)}$ is determined from (2.4)), derive the completion times $C_{(i,j)}$ of all jobs for positions $k < J$ with constraints (2.2). Rightshifting in the semiactive schedule implies that constraints (2.2) become equalities. For *ia-pb* we now have the result that it is sufficient to search over *sequences* when looking for an optimal *schedule*.

**Theorem 2.2** *For $[\alpha/fam, ia\text{-}pb, st_{g,i}, d_{(i,j)}/*]$, $\alpha \in \{1, F, ML1, ML, P\}$, there is an optimal semiactive schedule.*

**Proof:** The theorem is proven by contradiction. Suppose there is an optimal schedule which is not semiactive, i.e. at least one job can be locally rightshifted. For *ia-pb*, setup costs depend on the sequence of groups, regardless if a batch contains idle time or not. So setup costs are directly determined by $\pi$ (or by $\pi$ and $MA_k$ for $\alpha = P$), and will *not* change rightshifting a job. But earliness costs are not increased if we locally rightshift one job. Thus, there is always an optimal schedule which is semiactive.  □

Note that this result does not hold for batching types *ia-npb* and *ba*. In Figure 2.3 the *ia-pb* schedules $\sigma_I$ and $\sigma_V$ are semiactive. By Theorem 2.2, even for the multi-machine problems the search space for *ia-pb* is described by one sequence $\pi$ for $\alpha = ML$, or by $\pi$ and $MA_k$ for $\alpha = P$.

Similar to the leftshift dominance rule in Sprecher [116], Theorem 2.3 extends a single-machine property to multi-machines.

**Theorem 2.3** *If there is a solution for $[\alpha/fam,ia\text{-}pb,st_{g,i},d_{(i,j)}/*]$, $\alpha \in \{ML, P\}$, then there exists a sequence $\pi$ where the completion times in the corresponding semiactive schedule $\sigma$ increase monotonically, i.e. $C_{(i_{[k-1]},j_{[k-1]})} \leq C_{(i_{[k]},j_{[k]})}$, $k = 1, \ldots, J$, and $\sigma$ is optimal.*

**Proof:** As there is a solution for $[\alpha/fam,ia\text{-}pb,st_{g,i},d_{(i,j)}/*]$, $\alpha \in \{ML, P\}$, there is also an optimal one. Now, one can order the completion times of the optimal solution in nondecreasing order, with $\pi$ as the corresponding sequence. Consequently, there is an optimal $\pi$ where completion times are in nondecreasing order. $\hfill\square$

Theorem 2.3 leads to a simple dominance rule for multi-machine models in Chapter 5: a sequence is dominated if a job can be completed later in another sequence without altering the completion time of any other job. The solutions $\pi$ and $\sigma$ in Table 2.10 and Table 2.11 are consistent with Theorem 2.3: completion times increase monotonically, though jobs complete on *different* machines.

If only setup costs or only earliness costs are to be minimized, then the following properties of schedules can be derived.

**Theorem 2.4 (Minimize setup costs)** *For $[1/fam,\beta_1,st_{g,i},d_{(i,j)}/\sum sc_{g,i}]$, $\beta_1 \in \{ia\text{-}pb,$ $ia\text{-}npb, ba\}$, there exists an optimal schedule which is one block.*

**Proof:** Recall from Definition 2.4 that a block has no inserted idle time. Consider an optimal schedule $\sigma$: if in $\sigma$ no job can be leftshifted, we are done since $\sigma$ is one block. Otherwise, perform successive local leftshift operations until $\sigma$ is one block. With zero earliness weights, leftshifting neither increases earliness costs nor setup costs for all batching types. For zero earliness weights there is thus an optimal $\sigma$ which is one block. $\hfill\square$

By the same leftshift argument, each *ia-pb* schedule can be converted into an *ia-npb* schedule: simply leftshift all jobs in a group in *ia-pb* until they form a block with the first job in a group, resulting in an *ia-npb* schedule.[16] Consequently, feasibility of *ia-pb* implies feasibility of *ia-npb* (but note that this is not true for *ia-npb* and *ba*). A distinction between

---

[16]Consider e.g. the *ia-pb* schedule $\sigma_I$ in Figure 2.3: leftshifting $(3, 2)$ and $(3, 3)$ to $(3, 1)$ as the first job in the group we obtain the *ia-npb* schedule $\sigma_{II}$.

*ia-pb* and *ia-npb* therefore only makes sense if holding costs are nonzero and leftshifting a job changes the objective.

If we only consider the minimization of earliness costs, the following result is available (cf. Park et al. [98]).

**Theorem 2.5 (Minimize earliness costs)** *For  [1/fam,  $\beta_1, st_{g,i}, d_{(i,j)}/\sum w_{(i,j)} C_{(i,j)}$],  $\beta_1 \in \{ia\text{-}pb, ia\text{-}npb, ba\}$, there is an optimal schedule which is active.*

**Proof:** The proof is again by contradiction. Suppose there is an optimal schedule which is not active, i.e. at least one job can be globally or locally rightshifted without leftshifting another one. As setup costs are zero, rightshifting the job neither increases setup nor earliness costs. Thus, there is also an optimal active schedule.                    □

The following definitions and theorems exploit the family structure of the BSP. We are able to state a partial ordering between jobs which drastically reduces the search space for the algorithms in Chapters 3 and 5. In fact, this is the most important property of the BSP to be exploited in solution procedures.

**Definition 2.8** *A sequence $\pi$ (schedule $\sigma$) is called an EDDWF sequence (schedule), if jobs of one family are sequenced (scheduled) in increasing order of their deadlines, i.e. for jobs $(i_{[k]}, j_{[k]}), (i_{[l]}, j_{[l]}), i_{[k]} = i_{[l]}$ and $k < l$ $(C_{(i_{[k]}, j_{[k]})} < C_{(i_{[l]}, j_{[l]})})$ we have $d_{(i_{[k]}, j_{[k]})} < d_{(i_{[l]}, j_{[l]})}$ (earliest deadline within families).*

Recall, that jobs are already labeled in EDDWF order, so for two jobs $(i,j)$, $(i, j+1)$ we have $d_{(i,j)} < d_{(i,j+1)}$ (cf. Remark 2.1). By Theorem 2.3 sequence and schedule can always be ordered in the same way so that there is always a corresponding EDDWF sequence for an EDDWF schedule. The next theorem applies to a *unit time BSP*, i.e. problem $[1/fam, *, st_{g,i}, d_{(i,j)}, p_{(i,j)} = 1/*]$. All processing times are equal to one ($p_{(i,j)} = 1$) denoted by an entry $\beta_4$. Recall that processing times are integer valued, so each BSP can be converted into a unit time BSP.[17]

---

[17] Each job can be represented as $p_{(i,j)}$ unit time jobs with corresponding deadlines, cf. also $p_{(i,j)}$ $d_{(i,j)}$ for BSPUT(..) in Figure 4.2, p. 102.

**Theorem 2.6** *Any solution $\sigma$ of $[1/fam,*,st_{g,i},d_{(i,j)},p_{(i,j)} = 1/*]$ can be converted into an EDDWF solution with the same costs.*

**Proof:** All jobs of one family have the same weights and processing times. In a sequence $\pi$, let $A, B, C$ denote parts of $\pi$ (consisting of several jobs), and $C_A, C_B, C_C$ ($p_A, p_B, p_C$) the completion (processing) times of the parts. Consider a schedule, that is not an EDDWF solution, i.e. $\sigma = (C_A, C_{(i,j_2)}, C_B, C_{(i,j_1)}, C_C)$. Thus $C_{(i,j_2)} < C_{(i,j_1)} \leq d_{(i,j_1)} < d_{(i,j_2)}$. The schedule $\sigma = (C_A, \tilde{C}_{(i,j_1)}, C_B, \tilde{C}_{(i,j_2)}, C_C)$ with $\tilde{C}_{(i,j_1)} = C_{(i,j_2)}$, $\tilde{C}_{(i,j_2)} = C_{(i,j_1)}$ has the same objective function value because $w_{(i,j_1)} = w_{(i,j_2)} = h_i$. The completion times of the parts $A, B, C$ do not change because $p_{(i,j_1)} = p_{(i,j_2)} = 1$. The procedure can be repeated until $\sigma$ is an EDDWF schedule, completing the proof. $\quad\square$

Thus, for the unit time BSP we only need to consider EDDWF schedules. Ordering the jobs in EDDWF is called *ordered batch scheduling problem* in Monma and Potts [92]. They propose a DP algorithm where the states are given by the *number* of jobs scheduled for each family, and the family $i$ and starting time $t$ of the last scheduled job. Their algorithm is polynomial in the number of jobs but exponential in the number of families $N$. Thus, for fixed $N$ the algorithm is polynomial also for sequence dependent setups (there are $N^2$ different setups). But even for a fixed $N$ the state space is large because $t$ is a state variable. As we are primarily focusing on a small number of families and a large number of jobs, results for fixed $N$ are also interesting. The approach of Monma and Potts is the basis for the DP algorithm in Section 3.2.2.

Early completion of a job $(i, j)$ incurs inventory of family $i$ between $C_{(i,j)}$ and $d_{(i,j)}$. We refer to the production start for a family $i$ as the start time of the first job of the batch (note that we cannot use the start time of the *setup* because setups are sequence dependent in general). For the BSP we can now introduce *regenerative schedules*. The idea is similar to a property found by Wagner and Whitin [123] for the uncapacitated dynamic lotsizing problem.

**Definition 2.9** *A schedule $\sigma$ is called regenerative, if production for a family $i$ is started only if there is no inventory for $i$, i.e. if $d_{(i,j-1)} \geq C_{(i,j)} - p_{(i,j)}$ then jobs $(i, j-1)$ and $(i, j)$ are in the same batch.*

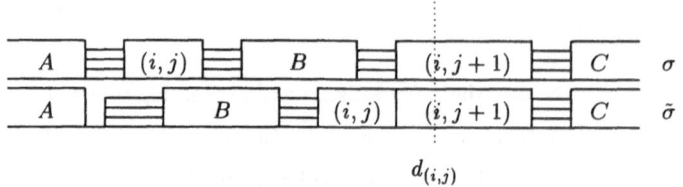

**Figure 2.10:** *Regenerative Schedule*

Regenerative schedules are considered in other problem settings as well; e.g. by Vickson et al. [122] and Julien and Magazine [73] (who call them "normal" schedules). As in the Wagner-Whitin problem, the following theorem states that for some cases we can restrict our attention to regenerative schedules. Unfortunately, the result does not extend to *ba*.

**Theorem 2.7** *If $\sigma$ is a feasible solution of $[\alpha/fam,\beta_1,st_{g,i},d_{(i,j)}/]$, $\alpha \in \{1, ML1\}$, $\beta_1 \in \{ia\text{-}pb,ia\text{-}npb\}$ then there is also a feasible solution $\tilde{\sigma}$ which is regenerative.*

**Proof:** Consider a schedule, which is not regenerative, i.e. $\sigma = (C_A, C_{(i,j)}, C_B, C_{(i,j+1)}, C_C)$. The jobs $(i,j)$ and $(i,j+1)$ are not in one batch though $C_{(i,j+1)} - p_{(i,j+1)} \leq d_{(i,j)}$. Then, for $\alpha = ML1$, a successor job $(u_i,j)$ of $(i,j)$ cannot be in $B$ (because then $C_{(i,j+1)} - p_{(i,j+1)} > \tilde{d}_{(i,j)}$ would hold).

Consider a schedule $\tilde{\sigma} = (C_A, \tilde{C}_B, \tilde{C}_{(i,j)}, C_{(i,j+1)}, C_C)$ with $\tilde{C}_{(i,j)} = C_{(i,j+1)} - p_{(i,j+1)}$ and $\tilde{C}_B = C_B - p_{(i,j)}$, where $(i,j)$ and $B$ are interchanged. $\tilde{\sigma}$ is feasible because $\tilde{C}_{(i,j)} \leq d_{(i,j)}$. By leftshifting $B$, each job completes before its deadline, and we do not violate precedence constraints for $\alpha = ML1$. Furthermore, due to the triangle inequality the setup time $A \to B$ is less or equal the setup time $A \to i \to B$. Thus, $B$ can be leftshifted $p_{(i,j)}$ time units without affecting $C_A$. $(i,j)$ and $(i,j+1)$ are in one batch which proves the theorem. $\square$

An illustration for the construction of regenerative schedules is given in Figure 2.10. Interchanging $(i,j)$ and $B$, we obtain the regenerative schedule $\tilde{\sigma}$ from $\sigma$. But the proof does not work for *ba*: $\sigma$ is feasible, but $\tilde{\sigma}$ not, $(i,j)$ is not available before $d_{(i,j)}$. In Figure 2.3, $\sigma_V$ is a non-regenerative schedule (but it is a semiactive one), job $(1,2)$ starts a batch though there is still inventory for family 1.

It is easy to see that a regenerative schedule implies an EDDWF schedule, but the inverse is not true. Thus, a simple consequence of Theorem 2.7 is that for feasibility we only need to consider EDDWF schedules (cf. also Monma and Potts [92] and Unal and Kiran [121]), a result that also holds for $ba$. The following theorem focuses on the optimization problem where holding costs are equal for all families, i.e. $h_i = h \; \forall i$.

**Theorem 2.8** *If $\sigma$ is an optimal solution of $[1/fam,\beta_1,st_{g,i},d_{(i,j)}/*,h_i = h]$, $\beta_1 \in \{ia\text{-}pb,ia\text{-}npb\}$ then there is also an optimal $\tilde{\sigma}$ which is regenerative.*

**Proof:** As in the proof of Theorem 2.7 we have to consider the cost contribution if $(i,j)$ and $B$ are interchanged.

W.l.o.g. let $h_i = 1 \; \forall i$, then $p_{(i,j)} = w_{(i,j)}$. Let $Z_{BSP}(\sigma)$ $(Z_{BSP}(\tilde{\sigma}))$ denote the objective function value of $\sigma$ $(\tilde{\sigma})$. Let $i^B$ $(i^A)$ be the family to which the first (last) job in part $B$ $(A)$ belongs.

For part $B$, which is leftshifted, we have $w_B \leq p_B$ because processing time in part $B$ is at most $p_B$, but $B$ may contain setups as well. Interchanging $B$ and $(i,j)$ the objective changes as follows:

$$Z_{BSP}(\tilde{\sigma}) = Z_{BSP}(\sigma) - w_B(\tilde{C}_B - C_B) - w_{(i,j)}(\tilde{C}_{(i,j)} - C_{(i,j)}) - sc_{i^A,i} - sc_{i,i^B} + sc_{i^A,i^B}$$

$$\overset{(i)}{\leq} Z_{BSP}(\sigma) + w_B p_{(i,j)} - w_{(i,j)} p_B \overset{(ii)}{\leq} Z_{BSP}(\sigma) + p_B p_{(i,j)} - p_{(i,j)} p_B = Z_{BSP}(\sigma)$$

Due to the triangle inequality, setup costs and setup times in $\sigma$ are not larger than in $\tilde{\sigma}$, i.e. $-sc_{i^A,i} - sc_{i,i^B} + sc_{i^A,i^B} \leq 0$, which explains $(i)$. We leftshift $B$ by $p_{(i,j)}$ and rightshift $(i,j)$ by $p_B$ with $w_B \leq p_B$, which explains $(ii)$. Thus $Z_{BSP}(\tilde{\sigma}) \leq Z_{BSP}(\sigma)$, which proves the theorem. □

Considering regenerative schedules, we again achieve a considerable reduction of the search space. Regenerative schedules are also a reasonable assumption in practice, therefore we present computational results in Chapter 3 with $h_i = 1 \; \forall i$ to exploit the family structure of the BSP. But even if this assumption is relaxed regenerative schedules are near optimal, cf. the results in Section 4.3.3, where instances with different holding costs are solved. Unfortunately, we cannot restrict ourselves to regenerative schedules and even not to EDDWF schedules in the multi-machine case:

For $\alpha = P$ there exists a counterexample where only a non-EDDWF schedule with overlapping jobs of one family (and therefore, a non-regenerative schedule) is feasible. For $\alpha = ML$, there is a counterexample where batches must be split to guarantee feasibility. Hence, also in this case only a non-regenerative schedule is feasible.

For the batching types *ia-npb* and *ba*, the next definition allows to discard schedules which have certain "anomalies".

**Definition 2.10** *For a given sequence $\pi$ in $[1/fam,\beta_1,st_{g,i},d_{(i,j)}/]$, $\beta_1 \in \{ia\text{-}npb,ba\}$, a schedule $\sigma$ is called late if $\sigma$ has the minimum makespan in $[1/fam,\beta_1,st_{g,i},r_{(i,j)}D/]$ (the corresponding mirror problem).*

If a schedule is *non*-late, at least the first job (at position $k = 1$) can be rightshifted if jobs are batched in a different way (maintaining the sequence). For *ia-pb*, the semiactive is also the late schedule, so a distinction between late and non late schedules makes no sense for *ia-pb*. It is straightforward, that among the feasible schedules there is a late one, but the *optimal* schedule is not necessarily late. How to derive a late schedule for a given sequence is discussed in Section 3.4.

In Figure 2.3 $\sigma_{IV}$ is a schedule for *ba* which is non-late, the *late* schedule for this sequence is $\sigma_{III}$ (where the first part of $\sigma_{IV}$ is rightshifted). But $\sigma_{III}$ incurs higher setup costs than $\sigma_{IV}$ and, e.g. for $h_i = 0$, $\sigma_{III}$ is *not* optimal. The construction of a non-late schedule for *ia-npb* is more difficult: assume $st_{0,3} = 4$ for $\sigma_{III}$. Then, in a non-late schedule job $(3,1)$ would complete at time 9, the start of the setup to job $(3,2)$. For $st_{0,3} = 4$ the late schedule would be $\sigma_{II}$.

As there is at least a *feasible* late schedule (if a feasible schedule exists), we will restrict ourselves to late schedules for *ia-npb* and *ba*.

## Results for Sequence Independent Setups and *ia-npb* and *ba*

Sequence independent setups allow to derive some more structural properties. Once we have partitioned the set of jobs into batches, a sequence independent setup can be associated with each batch. We can treat a *batch* as an entity and assign processing time, deadline and weight to a batch. We summarize some results which have been stated as well by (in chronological order) Mason and Anderson [90], Unal and Kiran [121] and Webster and Baker [124].

The time at which batch must complete such that all jobs in the batch become available before their deadline is called *batch deadline* $d_{(i,b)}^b$. The batch deadline depends on the batching type. More formally: let $\{(i, j_b), (i, j_b+1), \ldots, (i, j_{b+1}-1)\}, 1 \leq j_b \leq j_{b+1}-1 \leq n_i$ be the set of jobs which form the $b$-th batch of family $i$, then:

$$d_{(i,b)}^b = d_{(i,j_b)} \qquad\qquad\qquad\qquad \text{for } ba \qquad (2.16)$$

$$d_{(i,b)}^b = d_{(i,j_b)} + \sum_{j=j_b+1}^{j_{b+1}-1} p_{(i,j)} \qquad\qquad \text{for } ia\text{-}npb \qquad (2.17)$$

For $ba$, $d_{(i,b)}^b$ equals the deadline of the first job in the batch. For $ia\text{-}npb$, the first job completes at its deadline if the batch completes at $d_{(i,b)}^b$.

We refer to the ratio of the job weights and the batch processing time (with the sequence independent setup) as the *batch weight* $w_{(i,b)}^b$:

$$w_{(i,b)}^b = \frac{\sum_{j=j_b}^{j=j_{b+1}-1} w_{(i,j)}}{st_i + \sum_{j=j_b}^{j=j_{b+1}-1} p_{(i,j)}} \qquad\qquad (2.18)$$

For $h_i = 1$, and thus $w_{(i,j)} = p_{(i,j)}$, we can interpret the batch weight $w_{(i,b)}^b \in [0;1]$ as a *specific* weight for the fraction of time in a batch which is spent for production rather than setup. We are now able to state the following two results:

**Theorem 2.9** *If there is a feasible schedule $\sigma$ for $[1/fam,\beta_1,st_i,d_{(i,j)}/]$, $\beta_1 \in \{ia\text{-}npb,ba\}$ then there is a solution where batches are scheduled in nondecreasing order of the batch deadlines.*

**Proof:** The primary clue is that batches can now be seen as independent jobs because setups are sequence independent. In the mirror problem $[1/r_j/C_{max}]$ jobs must be scheduled in ERD order. For each problem with deadlines there is a mirror problem with release dates. Thus, there is a feasible schedule where batches (=jobs) are scheduled in EDD order, which proves the theorem. □

Consequently, for a given batching we can check feasibility easily.

The next result extends the Smith-Rule[18] to family scheduling. Two cases are easy: if there is only one family, the Smith-Rule can be applied directly; if there is only one job per family it must be applied on the family level (each family is a batch). For the intermediate case (which we are interested in) we can only state a much weaker property, namely that consecutively scheduled "early" batches are scheduled in nondecreasing order of batch weights.

**Theorem 2.10** *Let $C^b_{(i,b)}$ denote the completion time of batch $(i,b)$. If $C^b_{(i_1,b_1)} < C^b_{(i_2,b_2)} \leq d^b_{(i_1,b_1)} < d^b_{(i_2,b_2)}$, then there is an optimal solution for problem $[1/fam,\beta_1,st_i,d_{(i,j)}/*]$, $\beta_1 \in\{ia\text{-}npb,ba\}$ with the batch weights $w^b_{(i_1,b_1)} \leq w^b_{(i_2,b_2)}$.*

**Proof:** Both batches can be interchanged without altering the completion times of other batches. The (sequence independent) setup costs do not change since they are predetermined by the way jobs are batched. But earliness costs increase (with the same argument as in the Smith-Rule), which proves the theorem.                                   □

Theorems 2.9 and 2.10 are used for the genetic algorithm presented in Section 3.3.2.

## 2.7   Instance Generator

An instance generator for the weighted completion time problem without setups is described in Posner [103]. For zero setup times, it is known that scheduling the jobs in EDD order must be feasible (otherwise, there is no feasible solution). Thus, in Posner [103] deadlines are generated for each job such that the EDD schedule remains feasible (but is not necessarily the optimal one).

If setup times are nonzero, it is known from Section 2.5 that feasibility of a given instance can *not* be checked easily. Therefore, our generator is an extended version of the generators proposed by Unal and Kiran [121] and Woodruff and Spearman [125]: we first randomly generate a job sequence and construct a schedule as one block, taking the setups into account. Then, for each job a deadline is generated which does *not* make this schedule infeasible (but in general, another schedule will be optimal). In this way, even for

---

[18]Order all jobs in the order of shortest weighted processing time (SWPT), cf. e.g. Domschke et al. [39].

**Table 2.12:** *Input Parameters for the Instance Generator*

| | | |
|---|---|---|
| $\#b_i$ | : | Number of batches to be generated for family $i$ |
| $\#minjpb$ $\#maxjpb$ | : | Minimal and maximal number of jobs per batch |
| $minprc$ $maxprc$ | : | Minimal and maximal processing time of each job |
| $Z^-$ $Z^+$ | : | Tightness of deadlines, $0 \le Z^-$ , $Z^+ \le 1$ |
| $st_{g,i}$ | : | Setup time matrix |
| $\delta_{t2c}$ | : | Proportionality factor, $sc_{g,i} = \delta_{t2c} st_{g,i}$ |
| RASTER | : | The generator tries to assign deadlines to multiples of RASTER |

large and sequence dependent setups we are able to guarantee feasibility of an instance. Unal and Kiran [121] only consider the feasibility problem and thus do not generate cost parameters. Woodruff and Spearman [125], on the other hand, allow some (filler) jobs to be left unscheduled, and therefore consider an extended problem.

The input parameters for the generator are shown in Table 2.12. With $DU(a, b)$ we denote a discrete uniform distribution between $a$ and $b$. The generator is stated in pseudocode as follows:

$$\boxed{\text{Instance Generator}}$$

**Step 1:** Choose randomly a sequence of batches with $\#b_i$ batches for each family.

**Step 2:** Choose for each batch the batch size out of $DU(\#minjpb, \#maxjpb)$ and for each job the processing time $p_{(i,j)}$ out of $DU(minprc, maxprc)$. Construct a schedule without idle time. Set $T$ equal to the completion time of the last job of the schedule.

**Step 3:** Assign to each job $(i,j)$ of the schedule a deadline $d_{(i,j)}$ as follows:

1.   $I_{neg} = \max\{d_{(i,j-1)} + p_{(i,j)}; C_{(i,j)} - T \cdot Z^-\}$

2.   $I_{pos} = \max\{I_{neg}; C_{(i,j)}\} + T \cdot Z^+$

3.   Choose a deadline $d_{(i,j)}$ out of $DU(I_{neg}, I_{pos})$.

4.  Raster $d_{(i,j)}$ to the next multiple of RASTER.

5.  If $d_{(i,j)} < C_{(i,j)}$ for item availability, or $d_{(i,j)} < C_{(i,b)}^b$ for batch availability, assign $d_{(i,j)} = C_{(i,j)}$ or $d_{(i,j)} = C_{(i,b)}^b$, respectively.

In Step 1 a sequence with $\#b_i$ batches for each family is randomly generated; consecutive batches do not belong to the same family $i$.[19] For multi-level problems, ($\alpha = ML1, ML$), the batch sequence is generated as to be consistent with the (convergent) product structure.[20]

In Step 2 we randomly choose for each batch the number of jobs, and for each job a processing time. For multi-level problems the number of jobs in a batch is restricted by the number of predecessor jobs generated so far.[21]

With a given job sequence $\pi$ and $p_{(i,j)}$ for each job we construct a schedule as one block for the single-machine cases. For $\alpha = ML$ we leftshift each job on the appropriate machine as far as possible. In the parallel machine case, we "merge" $M$ single-machine instances to one $M$–machine instance.[22] Cost parameters are generated as follows: we normalize holding costs to one, thus $w_{(i,j)} = p_{(i,j)}$ and setup costs are proportional to setup times with a factor $\delta_{t2c}$ ("time to costs") such that $sc_{g,i} = \delta_{t2c} st_{g,i}$.

In Step 3 we assign deadlines to the individual jobs. This step is more complicated than in the generators of [121] and [125], and we depict this step in Figure 2.11: for each job a deadline is chosen at random out of the interval $[I_{neg}, I_{pos}]$. Deadlines are chosen according to Remark 2.1, consequently $d_{(i,j-1)} + p_{(i,j)} \leq I_{neg}$. For $I_{pos}$, the value $Z^+ \geq 0$ controls the tightness of deadlines with respect to the total length $T$ of the initial schedule. Then, $d_{(i,j)}$ is "rastered" in Step 3.4. In this way, some jobs of different families have identical

---

[19]Consequently, not every $\#b_i$ vector is feasible (which leads to a problem only if $N$ is small, e.g. $N = 4$). In the generators proposed in [121] and [125] the *total* number of batches is specified. But it is not guaranteed that the instances have a positive number of batches for each family. Small problems with a large number of families $N$ and a small number of jobs $J$ cannot be generated as proposed in [121] and [125].

[20]Cf. Figure 2.5 for an illustration: families 4 and 5 go into family 3, thus, at least one batch of 4 and 5 must precede a batch of family 3.

[21]And we cannot guarantee that the batch size is out of $DU(\#minjpb, \#maxjpb)$.

[22]Consequently, now deadlines may "interfere" and there is no longer $d_{(i,j-1)} < d_{(i,j)} - p_{(i,j)}$.

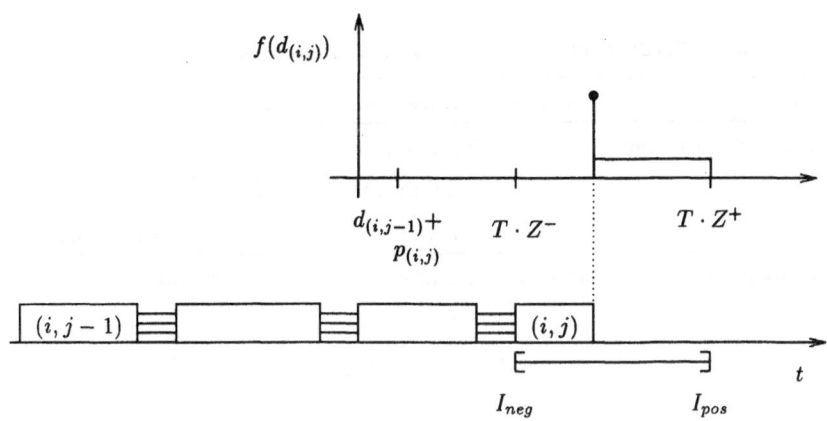

**Figure 2.11:** $f(d_{(i,j)})$ and $[I_{neg}; I_{pos}]$ for the Deadline Assignment

deadlines and an EDD ordering does not "propose" a solution. If $(i,j)$ is not yet available at $d_{(i,j)}$ (i.e. the deadline violates the feasible schedule), then we have either $d_{(i,j)} < C_{(i,j)}$ for *ia-pb* and *ia-npb*, or the deadline is smaller than the batch completion time, i.e. $d_{(i,j)} < C_{(i,b)}^b$ for *ba*. We then correct $d_{(i,j)}$ "upwards" in Step 3.5 to guarantee feasibility.

Using a value $Z^- > 0$ we enlarge the interval $[I_{neg}, I_{pos}]$ to the left and obtain $I_{neg} < C_{(i,j)}$ in general. This way we increase the probability that $d_{(i,j)} = C_{(i,j)}$ or $C_{(i,b)}^b$. For a large value $Z^- > 0$ the probability distribution $f(d_{(i,j)})$ shown in Figure 2.11 has a peak at the job (or batch) completion time; in this way we try to generate instances where some deadlines are "tight" even if capacity utilization is low.

For the computational results in Chapter 3 we choose different *factors* for the **characterization of instances**. The factors are summarized in Table 2.13: the parameters for the generator are set accordingly.

Instances differ in the **capacity utilization** $\rho$. $\rho$ does not only depend on the tightness of deadlines $Z^-$ and $Z^+$, but also on $(\#minjpb, \#maxjpb)$. For the generation of "general" instances without a certain $\rho$ we choose $\#minjpb = \#maxjpb = 1$ and $(Z^-, Z^+) = (0.05, 0.01)$. These instances are supposed to have no special structure. The generated schedule contains much setup time, but a solution, where jobs are batched, may utilize less

**Table 2.13:** *Factors, Values and Parameters of the Generator*

| Factor | Symbol | Values | Generator |
|--------|--------|--------|-----------|
| capacity utilization | $\rho$ | high (H) | $(Z^-, Z^+) = (0.05, 0.01)$ |
| | | medium (M) | $(Z^-, Z^+) = (0.05, 0.1)$ |
| | | low (L) | $(Z^-, Z^+) = (0.1, 0.2)$ |
| setup structure | *st-struc* | group ($gp$) | cf. Appendix A |
| | | random ($rd$) | |
| | | sequence ($sq$) | |
| | | sequence independent ($si$) | |
| setup significance | $\theta$ | low ($l$) | $\theta = 0.5$ |
| | | high ($h$) | $\theta = 1$ |

capacity. On the other hand, if we want to evaluate the influence of $\rho$ on the algorithmic performance, we set $(\#minjpb, \#maxjpb) = (1, 2)$, and choose $(Z^-, Z^+)$ as in Table 2.13.

Furthermore, we distinguish different **setup structures** (for a similar idea cf. Fleischmann [50]), characterizing the setup matrix. We distinguish between

- group structure ($gp$): families can be clustered into groups, where setups between families of the same group are smaller than setups between different groups. For instances with $N = 6$, the families $(1, 2)$, $(3, 4)$ and $(5, 6)$ form 3 setup groups. With ($gp$) we imitate the case of major and minor setups.

- random structure ($rd$): setups are random between families, but the triangle inequality holds true. With ($rd$) the setups are supposed to have no special structure.

- sequence structure ($sq$): setups $g \to i$ between families $g$ and $i$ are large for $g < i$ and small for $g > i$. The setup structure ($sq$) is motivated by an example from the color sequence: setups from light to a dark color are small but the inverse order incurs large setups.

- sequence independent setup structure ($si$): setups are sequence independent.

The different setup matrices ($st_{g,i}$) are given in Appendix A.

The last factor to be considered in the computational study is the **setup significance** $\theta$. $\theta$ is the ratio between average setup and processing time, defined as:

$$\theta = 2\frac{\sum_{g,i} st_{g,i}}{maxprc + minprc}.$$

For large $\theta$, the setup time is large compared to the average job processing time. In our experiments, we either use instances with a high ($\theta = 1$) or a low ($\theta = 0.5$) setup significance. By definition, we have $\theta = 0$ for zero setup times.

# Chapter 3

# Methods for the Single-Machine Case

This chapter is dedicated to the development of algorithms for the single-machine case ($\alpha = 1$). Algorithms differ with respect to batching types and sequence dependent or independent setups. In Section 3.1, we describe the basic enumeration scheme which is used by the branch and bound (B&B) algorithms for the batching types *ia-pb*, *ia-npb* and *ba*. Section 3.2 deals with *ia-pb*, for which we present two exact and one heuristic algorithm and study their performance at the end of the section. Section 3.3 is concerned with *ia-npb*; in the B&B algorithm we must derive a minimum cost schedule from a job sequence (cf. Section 3.3.1.1), and dominance rules become more complicated than for *ia-pb* (cf. Section 3.3.1.2). For sequence independent setups a genetic algorithm handles batching type *ia-npb* (cf. Section 3.3.2) as well as *ba* (cf. Section 3.4.2). On the other hand, for *ba* in Section 3.4 the B&B algorithm again requires modification. We derive a late schedule from a sequence in Section 3.4.1.1 and state the dominance rule for *ba* in Section 3.4.1.2. Computational experiments are presented at the end of Sections 3.3 and 3.4, respectively. Successively solving the single-machine case, the genetic algorithm can also be used for the flow-shop case, which we describe in Section 3.4.4.

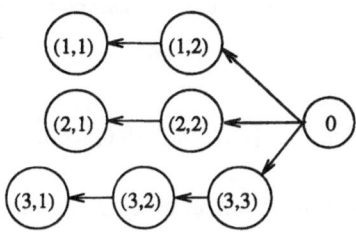

**Figure 3.1:** *EDDWF Precedence Graph*

# 3.1 Basic Enumeration Scheme

This section presents the common basis of the B&B algorithms described in Sections 3.2, 3.3 and 3.4. Jobs are sequenced backwards, i.e. at stage 1 a job is assigned to position $J$, at stage 2 to position $J - 1$, at stage $s$ to position $J - s + 1$[1]. We only examine EDDWF sequences (cf. Definition 2.8) as if there were precedence constraints between the jobs and therefore exploit the structural properties of the BSP. The precedence graph for the example in Figure 2.3 (p. 27) is shown in Figure 3.1. Via the EDDWF ordering, we decide in fact at each stage $s$ which *family* to schedule.

In the B&B algorithm, we build partial solutions for the BSP; precise definitions are given in the following.

**Definition 3.1** *An s-partial sequence $\pi^s$ assigns $s$ jobs to the last $s$ positions of sequence $\pi$, i.e.*

$$\pi^s = \left( (i_{[J-s+1]}, j_{[J-s+1]}), (i_{[J-s+2]}, j_{[J-s+2]}), \ldots, (i_{[J]}, j_{[J]}) \right).$$

$(i^s, j^s)$ *is used to denote the job* $(i_{[J-s+1]}, j_{[J-s+1]})$ *under consideration at stage $s$.*

Then, a $J$-partial sequence $\pi^J$ is a sequence as in Definition 2.1.

---

[1] If only the minimization of setup costs is considered, a *forward* sequencing scheme is more appropriate because there is an optimal schedule which forms a block (Theorem 2.4). Hence, we will not consider this objective in the B&B algorithms.

**Definition 3.2** *An s-partial schedule $\sigma^s$ assigns completion times to the last s positions of sequence $\pi$, i.e.*

$$\sigma^s = \left( C_{(i_{[J-s+1]},j_{[J-s+1]})}, C_{(i_{[J-s+2]},j_{[J-s+2]})}, \ldots, C_{(i_{[J]},j_{[J]})} \right)$$

*A partial schedule $\omega^s$ is called completion of $\sigma^s$ if $\omega^s$ extends $\sigma^s$ to a solution $\sigma$ and we write $\sigma = (\omega^s, \sigma^s)$.[2]*

Analogously, a $J$-partial schedule $\sigma^J$ is a schedule as in Definition 2.2.

A job is eligible at stage $s$ if all its (precedence graph based) predecessors are scheduled. An $s$-partial schedule is extended by scheduling an eligible job at stage $s+1$ of the search tree. If an $s$-partial schedule is bounded, or if it is dominated by another $s$-partial schedule, backtracking occurs. We apply depth-first search in our enumeration.

Each ($s$-partial) sequence $\pi^s$ uniquely defines a minimum cost ($s$-partial) schedule $\sigma^s$. This is the semiactive schedule for *ia-pb* (cf. Theorem 2.2), but for *ia-npb* the minimum cost schedule must be derived. Furthermore, we only consider late schedules in the enumeration (cf. Definition 2.10).[3] Enumerating is done over all sequences, stopping after all sequences have been (implicitly) examined, the best solution found is optimal.

The number $N_{all}$ of all EDDWF sequences is

$$N_{all} = \frac{J!}{\prod_{i=1}^{N} n_i!}$$

which is exponential in the number of jobs. Consequently, bounding and dominance rules are needed to restrict the search; the development of effective rules will be our main concern in this chapter.

Table 3.1 depicts attributes of $s$-partial schedules. For each scheduling stage $s$, we identify the job $(i^s, j^s)$ under consideration, the start time $t(\sigma^s)$ and the cost $c(\sigma^s)$ of the $s$-partial schedule. The set of currently unscheduled jobs is denoted by $\mathcal{US}^s$, or alternatively, by

---

[2]With this notation we refer to the concatenation of two vectors, and not to the mathematical definition where $(\omega^s, \sigma^s)$ means a pair of vectors.

[3]Especially for *ba* this may be suboptimal if a non-late schedule has lower costs.

**Table 3.1:** *Attributes of Partial Schedules I*

| | |
|---|---|
| $(i^s, j^s)$ | job under consideration at stage $s$ |
| $UB$ | upper bound, cost of the current best solution |
| $t(\sigma^s)$ | start time of $\sigma^s$, i.e. $C_{(i^s,j^s)} - p_{(i^s,j^s)}$ |
| $c(\sigma^s)$ | cost of $\sigma^s$ without the setup for $(i^s, j^s)$ |
| $\mathcal{AS}^s(\mathcal{US}^s)$ | set of jobs already scheduled (unscheduled) in the $s$-partial schedule $\sigma^s$ |
| $v_i$ | number of jobs of family $i$ currently unscheduled |
| $\mathcal{UI}^s$ | set of families to which jobs in $\mathcal{US}^s$ belong to |

the vector $(v_1, \ldots, v_i, \ldots, v_N)$ (with EDDWF ordering). $\mathcal{UI}^s$ denotes to which families the jobs in $\mathcal{US}^s$ belong to, $UB$ is the current upper bound.

Two simple bounds, which hold for all batching types, can now be derived:
The **feasibility bound** in Theorem 3.1 states that for a given $\sigma^s$, all currently unscheduled jobs in $\mathcal{US}^s$ must be scheduled between time zero and the start time $t(\sigma^s)$, and, furthermore, we need a setup time for each family in $\mathcal{UI}^s$.

**Theorem 3.1 (Feasibility bound)** *A lower bound* $\mathsf{T}^s$ *for the time needed to schedule a completion* $\omega^s$ *of* $\sigma^s$ *between time zero and* $t(\sigma^s)$ *is*

$$\mathsf{T}^s = \sum_{i \in \mathcal{UI}^s} \min_{\substack{g=0,\ldots,N \\ g \neq i}} \{st_{g,i}\} + \sum_{(i,j) \in \mathcal{US}^s} p_{(i,j)}.$$

$\sigma^s$ *has no feasible completion* $\omega^s$ *if* $t(\sigma^s) - \mathsf{T}^s < 0$.

**Proof:** Obvious.                                                       □

The **cost bound** is stated in Theorem 3.2. The costs $c(\sigma^s)$ of an $s$-partial schedule are a lower bound for all extensions of $\sigma^s$, and for any completion $\omega^s$ at least one setup for each family in $\mathcal{UI}^s$ must be performed. Together, we obtain a (rather weak) lower bound for the costs of $(\omega^s, \sigma^s)$.

**Theorem 3.2 (Cost bound)** *A lower bound $C^s$ for the minimum cost to schedule a completion $\omega^s$ of $\sigma^s$ between time zero and $t(\sigma^s)$ is*

$$C^s = \sum_{i \in \mathcal{UI}^s} \min_{\substack{g=0,\dots,N \\ g \neq i}} \{sc_{g,i}\}.$$

$\sigma^s$ *cannot be extended to an improved solution* $\sigma^J = (\omega^s, \sigma^s)$ *if* $C^s + c(\sigma^s) \geq UB$.

**Proof:** Obvious.                                                                                □

Feasibility and cost bound are checked for each $s$-partial schedule $\sigma^s$. Clearly, $T^s$ and $C^s$ can be easily updated during the search.

The **dominance rules** of the B&B algorithms compare two $s$-partial schedules $\sigma^s$ and $\overline{\sigma}^s$, which schedule the same set of jobs, such that $\mathcal{AS}^s = \overline{\mathcal{AS}}^s$. In this notation, $\overline{\sigma}^s$ denotes the $s$-partial schedule currently under consideration while $\sigma^s$ denotes a previously enumerated schedule which may dominate $\overline{\sigma}^s$. During the enumeration, we store attributes $t(\sigma^s)$ and $c(\sigma^s)$ of partial schedules for each job set $\mathcal{AS}^s$ and family $i^s$.[4] Then, at any point in the enumeration, either new information is stored for $\mathcal{AS}^s$, or we can compare the attributes $t(\overline{\sigma}^s), t(\sigma^s)$ and $c(\overline{\sigma}^s), c(\sigma^s)$ of $\sigma^s$ and $\overline{\sigma}^s$ (and may dominate $\overline{\sigma}^s$). Different batching types require different dominance rules, so the rules are stated separately in the next sections.

The number of $s$-partial schedules is exponential in the problem size so that also storage requirements for the dominance rules grow exponentially. Therefore, for larger instances, we cannot store the information for all pairs $(\mathcal{AS}^s, i^s)$ (*maximal storage size*) but specify a smaller *memory limit* which can be handled by the computer. The idea is that only a part of the maximal storage size is needed for two reasons: first, if partial schedules are dominated in the subsequent enumeration process we do not need to store the information corresponding to them. Second, if we encounter a "better" schedule, we "overwrite" the existing information and thus do not need additional storage. However, when the prespecified memory limit is "filled" with information, the memory limit is exceeded if enumeration reaches an $s$-partial schedule $\sigma^s_{new}$ and no information is stored for the corresponding job

---

[4]The implementation uses ideas of Baker and Schrage [9].

set $\mathcal{AS}^s$.[5]

If only regenerative schedules need to be considered (cf. Theorem 2.8), we employ a **branching rule** as follows: scheduling $(i^s, j^s)$ at stage $s$, job $(i^s, j^s - 1)$ becomes eligible. If $t(\sigma^s) \leq d_{(i^s, j^s - 1)}$, we extend $\sigma^s$ only with $(i^s, j^s - 1)$ (i.e. we batch $(i^s, j^s)$ and $(i^s, j^s - 1)$) and do not consider any other job as an extension of $\sigma^s$. We do not need to enumerate partial schedules where $\sigma^s$ is extended with a job $(g, j)$ with $g \neq i^s$, because then, the resulting schedule is not regenerative. E.g. in Figure 3.3, $\overline{\sigma}^5$ must be extended with job $(2, 1)$ because $d_{(2,1)} = 9 \geq t(\overline{\sigma^5}) = 7$; any other extension of $\overline{\sigma}^5$ leads to a non-regenerative schedule.

## 3.2 Item Availability – Preemptive Batching

In this section we consider the problem $[1/fam, st_{g,i}, ia\text{-}pb/\sum w_{(i,j)} C_{(i,j)} + \sum sc_{g,i}]$, written as $[1 \diamond ia\text{-}pb]$ for short. We develop a B&B, a DP and a construction and improvement algorithm, denoted as B&B$[1 \diamond ia\text{-}pb]$, DP$[1 \diamond ia\text{-}pb]$ and C&I$[1 \diamond ia\text{-}pb]$, respectively.

### 3.2.1 Branch&Bound

B&B$[1 \diamond ia\text{-}pb]$ enumerates all sequences as described in Section 3.1 and for a given sequence we construct the corresponding semiactive schedule (cf. Theorem 2.2).

In addition to the cost and feasibility bound described above we use an *occupation bound* (cf. Definition 2.6) for *ia-pb*. It schedules all jobs in $\mathcal{US}^s$ in EDD order without considering setups, checks feasibility and derives a lower bound on earliness costs. An algorithmic description is given in the following.

---

[5] We also implemented other storing schemes: $(i)$ information is stored only if $\mathcal{AS}^s$ is a subset $\mathcal{SM}$ of all jobs, and $\mathcal{SM}$ is determined such that the information of all $\mathcal{AS}^s \in \mathcal{SM}$ can be stored within the given memory limit. $(ii)$ we just continued enumeration though no new information could be stored. The computational experience was disappointing in both cases. But in general, algorithms tend to exceed a certain time limit first before exceeding the memory limit, i.e., if we encounter memory problems for instances with large $N$, we encounter running time problems, too. Hence, we do not examine any other improved storage scheme.

$$\boxed{\text{Occupation bound}}$$

Given $\sigma^s$, $t(\sigma^s)$, $c(\sigma^s)$, $UB$, $\mathcal{US}^s$, $\mathcal{UI}^s$ and $h^{min} = \min_{i=1,\dots,N}\{h_i\}$

**Step 1:** Determine the occupation of all jobs $(i,j) \in \mathcal{US}^s$, where the job $(i,j)$ with the largest deadline completes at $t(\sigma^s)$. Set $t^{occ}$ to the start of the occupation.

**Step 2:** Calculate $EC^{min}$ as the sum of the earliness costs of all jobs $(i,j) \in \mathcal{US}^s$, with $w_{(i,j)} = h^{min} p_{(i,j)}$.

**Step 3:** (*Extended feasibility bound*): $\sigma^s$ has no feasible extension if $t^{occ} < 0$

**Step 4:** (*Extended cost bound*): $\sigma^s$ has no extension leading to an improved solution if

$$c(\sigma^s) + \sum_{i \in \mathcal{UI}^s} \min_{\substack{g=0,\dots,N \\ g \neq i}} \{sc_{g,i}\} + EC^{min} \geq UB$$

For each $\sigma^s$, the occupation bound must be calculated at each node in the enumeration tree; we solve a relaxed problem where setups are omitted. The occupation bound derives a lower bound on the earliness costs as the earliness costs are calculated with $h_{min}$. Compared to Theorem 3.1, we obtain a tighter feasibility bound if the idle time in the occupation is larger than the minimum setup time, we then have $t^{occ} \leq t(\sigma^s) - T^s$. Furthermore, compared to Theorem 3.2, we have a tighter cost bound due to the additional term $EC^{min} > 0$.

The dominance rule for B&B[$1\diamond ia\text{-}pb$] compares attributes of two $s$-partial schedules $\sigma^s$ and $\overline{\sigma}^s$. The case $\overline{i^s} = i^s$ is easily understood: due to the EDDWF ordering, $\sigma^s$ and $\overline{\sigma}^s$ schedule the same job at stage $s$, i.e. $(i^s, j^s) = (\overline{i^s}, \overline{j^s})$. If $\overline{i^s} \neq i^s$, we make $\overline{\sigma}^s$ "comparable" with $\sigma^s$ with a setup from $\overline{i^s}$ to $i^s$.

**Theorem 3.3** *Consider two $s$-partial schedules $\sigma^s$ and $\overline{\sigma}^s$ with $\mathcal{AS}^s = \overline{\mathcal{AS}}^s$. $\sigma^s$ dominates $\overline{\sigma}^s$ if*

$\quad$ (*i*) $\quad t(\overline{\sigma}^s) + st_{\overline{i^s},i^s} \leq t(\sigma^s) \qquad$ *and*

$\quad$ (*ii*) $\quad c(\overline{\sigma}^s) - sc_{\overline{i^s},i^s} \geq c(\sigma^s)$.

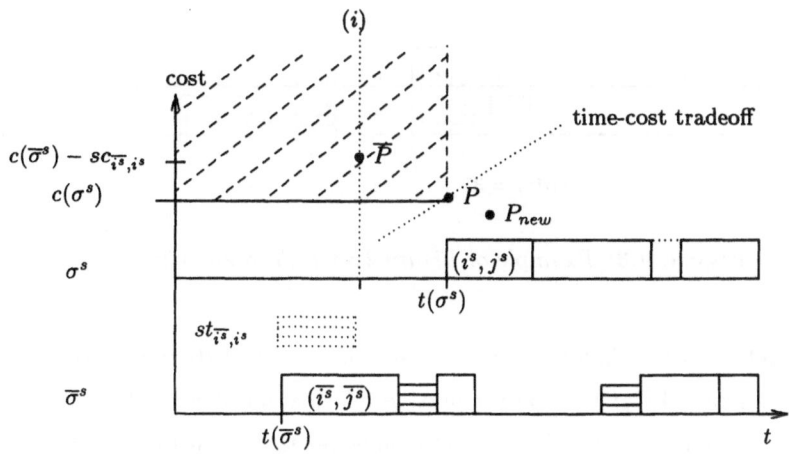

**Figure 3.2:** *Illustration of Theorem 3.3*

**Proof:** Let $i^\omega$ be the family of the (last) job in a completion $\omega^s$ of $\overline{\sigma}^s$. Then $st_{i^\omega,i^s} \leq st_{i^\omega,\overline{i^s}} + st_{\overline{i^s},i^s}$, analogously for setup costs due to the triangle inequality. Thus, any completion $\omega^s$ of $\overline{\sigma}^s$ is also a feasible completion of $\sigma^s$ because of $(i)$; if $(\omega^s, \overline{\sigma}^s)$ is feasible, $(\omega^s, \sigma^s)$ is feasible, too. Due to $(ii)$, for any $\omega^s$, the solution $(\omega^s, \sigma^s)$ has lower costs than $(\omega^s, \overline{\sigma}^s)$, so $\sigma^s$ dominates $\overline{\sigma}^s$, completing the proof. $\qquad\qquad\square$

Consider the illustration in Figure 3.2: in a time-cost diagram, we assign a point $P$ to schedule $\sigma^s$; $P$ represents the pair $t(\sigma^s), c(\sigma^s)$ stored for the job set $\mathcal{AS}^s$. $P$ is more "efficient" than all other points $\overline{P}$ of $\overline{\sigma}^s$ in the shaded region of the diagram, and $\sigma^s$ dominates $\overline{\sigma}^s$.

In addition, Figure 3.3 with the example of Figure 2.3 illustrates Theorem 3.3. $\sigma^5$ and $\overline{\sigma}^5$ schedule the same set of jobs $\mathcal{AS}^5 = \{(1,2),(2,2),(3,1),(3,2),(3,3)\}$ and $i^5 = 3 \neq \overline{i^5} = 2$. Theorem 3.3 is applied because $(i)$ $7+1 \leq 9$ and $(ii)$ $37-5 \geq 21$ so that $\sigma^5$ dominates $\overline{\sigma}^5$.

If enumeration reaches an $s$-partial schedule $\sigma^s_{new}$, we distinguish three cases: $(i)$ no information is stored for the pair $(\mathcal{AS}^s, i^s)$ yet. Then, we store $t(\sigma^s)$ and $c(\sigma^s)$ of the current schedule $\sigma^s_{new}$. $(ii)$ $\sigma^s_{new}$ is dominated and backtracking occurs. $(iii)$ $\sigma^s_{new}$ with $i^s = i^s_{new}$

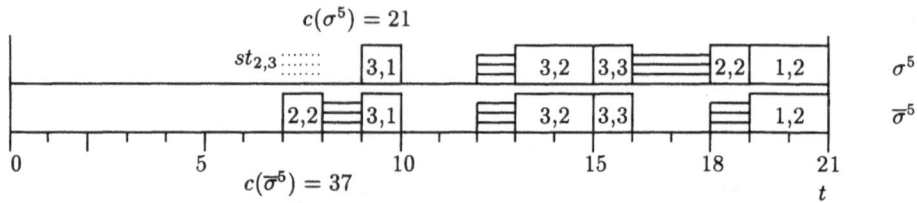

**Figure 3.3:** *Example of Figure 2.3 for Theorem 3.3:* $\sigma^5$ *dominates* $\bar{\sigma}^5$

is a "better" schedule for the job set $\mathcal{AS}^s$ than $\sigma^s$, and $t(\sigma^s)$ and $c(\sigma^s)$ are overwritten. For one pair $(\mathcal{AS}^s, i^s)$ we only store the information of *one* schedule $\sigma^s$ and apply the following "storing scheme": $\sigma_{new}^s$ must be better than $\sigma^s$ with respect to time *and* costs. We overwrite $t(\sigma^s)$ and $c(\sigma^s)$ if $c(\sigma_{new}^s) - \xi_{t2c}(t(\sigma_{new}^s) - t(\sigma^s)) \leq c(\sigma^s)$. For the time-cost tradeoff factor $\xi_{t2c}$, we use the sum of earliness weights of all unscheduled jobs, i.e. $\xi_{t2c} = \sum_{(i,j) \in \mathcal{US}^s} w_{(i,j)}$. $\xi_{t2c}$ is based upon the fact that in finding an optimal solution we must maintain feasibility. Thus, if $\sigma_{new}^s$ and $\sigma^s$ have the same costs, the schedule with the later start time is preferred (because it is more likely to have a feasible extension).[6] In Figure 3.2, $\sigma_{new}^s$ is better than $\sigma^s$ if the point $P_{new}$ is below the dotted time-cost tradeoff line.

### 3.2.2   Dynamic Programming

The same ideas used to partial schedules for job sets are also employed in the DP algorithm DP[1◇ia-pb]. Similar to backward *sequencing* in B&B, we employ backward *recursion* in DP[1◇ia-pb]. Jobs are in EDDWF order, and the vector $(v_1, \ldots, v_i, \ldots, v_N)$ of currently unscheduled jobs identifies $\mathcal{US}^s$. Let $f(v_1, \ldots, v_i, \ldots, v_N, i, t)$ denote the minimum cost of scheduling a certain set of jobs, with the last scheduled job from family $i$ and time $t$ as the start time of the partial schedule (in B&B we would consider stage $s = \sum_{g=1}^{N}(n_g - v_g)$ with job $(i^s, j^s) = (i, v_i)$ and $t(\sigma^s) = t$). We can now define the following recurrence

---

[6]But the time-cost tradeoff also allows for the new schedule to be stored though $t(\sigma^s) > t(\sigma_{new}^s)$ if $c(\sigma_{new}^s) \ll c(\sigma^s)$. Then, there may be a feasible extension for $\sigma^s$ but none for $\sigma_{new}^s$. Nevertheless, this "storing scheme" turned out to be more efficient.

equation (3.1) between states:

$$f(v_1, \ldots, v_i - 1, \ldots, v_N, i, t') = \min_{g=1,\ldots,N} \{ f(v_1, \ldots, v_i, \ldots, v_N, g, t) +$$
$$+ sc_{i,g} + w_{(i,v_i)}(d_{(i,v_i)} - t_c) \} \tag{3.1}$$
$$t_c = \min\{d_{(i,v_i)}, t - st_{i,g}\}$$
$$t' = t_c - p_{(i,v_i)}$$
$$f(n_1, \ldots, n_i, \ldots, n_N, g, T) = 0; \ g = 1, \ldots, N \tag{3.2}$$

In the transition from state $(v_1, \ldots, v_i, \ldots, v_N, g, t)$ to $(v_1, \ldots, v_i-1, \ldots, v_N, i, t')$, $g \neq i$, we schedule job $(i, v_i)$ before job $(g, v_g)$, which starts at time $t$. The semiactive completion time of $(i, v_i)$ is $t_c = \min\{d_{(i,v_i)}, t - st_{i,g}\}$. Costs increase between states by $sc_{i,g} + w_{(i,v_i)}(d_{(i,v_i)} - t_c)$. The start time of $(i, v_i)$ is $t' = t_c - p_{(i,v_i)}$. The initialization of the DP algorithm is given in (3.2), $T$ denotes the maximum deadline of all jobs. The minimum cost $Z^*$ to schedule all jobs also includes the start setup and is given by

$$Z^* = \min_{i=1,\ldots,N} \{ sc_{0,i} + f(0, \ldots, 0, i, t') | t' - st_{0,i} \geq 0 \}$$

Monma and Potts [92] use a recurrence equation similar to (3.1) to show that if only EDDWF sequences are considered the problem is polynomially solvable for a fixed number of families $N$. In their formulation *"the large number of possible values for the state variable $t$ may lead to computer storage problems"*(cf. [92], p. 801). For *ia-pb*, we know that the semiactive schedule is optimal for a given sequence and thus time $t'$ depends on $t$. Equation (3.1) is stated such that $t'$ is a function of $t$, i.e. $t' = t'(t)$. $t'$ is calculated backward oriented so that only a small number of values for the state variable $t$ needs to be considered.

Further reductions of the state space can be achieved using the feasibility bound and the dominance rule of B&B[1◇*ia-pb*]. We can eliminate all states which do not have a feasible extension due to Theorem 3.1. More formally, we eliminate a state $(v_1, \ldots, v_i, \ldots, v_N, i, t')$ where

$$\sum_{i=1}^{N} \sum_{j=1}^{v_i} p_{(i,j)} + \sum_{i|v_i>0} \min_{\substack{g=0,\ldots,N \\ g \neq i}} st_{g,i} > t'.$$

The first term is the sum of processing times of unscheduled jobs, the second term calculates the minimum setup times. The cost bound, however, cannot be applied here because the first feasible solution is found in the last stage.

The dominance rule of Theorem 3.3 is used to eliminate *inefficient* states. Dumas et al. [45] use the same idea in their DP algorithm for the TSP with time windows. At each scheduling stage $s$ we compare states which schedule the same set of jobs $\mathcal{AS}^s$ and *"we only conserve the Pareto optimal elements"* (Dumas et al. [45], p. 368). State $(v_1, \ldots, v_i, \ldots, v_N, g, t')$ with costs $f$ dominates state $(v_1, \ldots, v_i, \ldots, v_N, i, \overline{t'})$ with costs $\overline{f}$ if

$$\overline{t'} + st_{i,g} \leq t', \text{ and } \overline{f}(v_1, \ldots, v_i, \ldots, v_N, i, \overline{t'}) - sc_{i,g} \geq f(v_1, \ldots, v_i, \ldots, v_N, g, t').$$

In the implementation of DP[1$\diamond$*ia-pb*] we check for dominance when a new state is created. Contrary to B&B[1$\diamond$*ia-pb*], the DP algorithm may store information of more than one partial schedule for the pair $(\mathcal{AS}^s, i^s)$ if more than one time-cost pair is Pareto optimal.

As we only need to consider regenerative schedules, we do not create states where the schedule would be non-regenerative (cf. the description in Section 3.1).

DP[1$\diamond$*ia-pb*] uses very much the same ideas as B&B[1$\diamond$*ia-pb*] but a different search strategy. DP first considers all schedules of one stage $s$ (similar to breadth-first search in B&B) and eliminates the "inefficient" ones, whereas B&B[1$\diamond$*ia-pb*] extends one $s$-partial schedule to an $s + 1$-partial schedule (depth-first search) where some schedules are dominated.

### 3.2.3 Construction and Improvement

The last algorithm we present for [1$\diamond$*ia-pb*] is a simple construction and improvement heuristic (C&I[1$\diamond$*ia-pb*]). In the construction step an initial schedule is generated which is then improved by "delete-and-insertion" moves. Ten Kate [74] and Woodruff and Spearman [125] describe similar heuristics. In [125], moves, deleting a job at one position and inserting at another one turned out to be more effective than "swap"-moves, where two jobs are interchanged. An algorithmic description of C&I[1◊*ia-pb*] is given in the following.

---

**Algorithm C&I[1◊*ia-pb*]**

---

**Step 1:** (*Construction of the initial sequence*)

Schedule all jobs in order of decreasing $d_{(i,j)}$ (taking into account setups and deadlines); $k := J$;

**Step 2:** (*Improvement I*)

Delete job $(i,j)$ at position $k$ of the initial sequence and insert it after job $(i, j-1)$ if the objective function value decreases; $k := k - 1$;

**Step 3:** if$(k = 1)$ then $k := J$; else GOTO **Step 2**;

**Step 4:** (*Improvement II*)

Delete job $(i,j)$ at position $k$ of the improved sequence and insert it at any position between $k$ and the position of job $(i, j-1)$ if the objective function value decreases. $k := k - 1$;

**Step 5:** if $(k = 1)$ then STOP; else GOTO **Step 4**;

---

The initial sequence orders all jobs in EDD order, and often, due to the setups, the initial sequence may be infeasible (start at a "negative" start time); also, it incurs penalty costs proportional to that start time. In Step 2, a job is moved forward and batched with the preceding job of its family if costs decrease. In the second improvement in Step 4, we test each position between job $(i,j)$ and $(i, j-1)$. In Step 4, $J$ jobs are tested at most $J$ times, so the complexity of C&I[1◊*ia-pb*] is bounded by $\mathcal{O}(J^2)$.

## 3.2.4 Computational Results

This section reports on computational experience with the different algorithms for [1◊*ia-pb*]. The effectiveness of the bounding and dominance rules for B&B is illustrated in Table 3.2. Figure 3.4 allows for an assessment of the problem sizes which can be solved to optimality. The influence of different setup-structures and capacity utilization is examined

**Table 3.2:** *Effectiveness of Bounding and Dominance Rules in B&B[1◇ia-pb]*

| Occupation Bound | Regenerative Schedules | Dominance Rule Theorem 3.3 | $(N, J)$ | | | |
|:---:|:---:|:---:|:---:|:---:|:---:|:---:|
| | | | $(3, 22)$ | | $(5, 18)$ | |
| | | | $R_{avg}$ | $R_{max}$ | $R_{avg}$ | $R_{max}$ |
| | | | 14.84 | 151.50 | 51.06 | 600.08 |
| + | | | 3.48 | 28.78 | 66.25 | *1200.00 |
| | + | | 1.25 | 14.19 | 4.80 | 38.38 |
| | | + | 0.06 | 0.24 | 0.16 | 0.55 |
| + | + | | 0.55 | 6.05 | 1.00 | 7.22 |
| | + | + | 0.04 | 0.14 | 0.10 | 0.36 |
| + | + | + | 0.05 | 0.19 | 0.12 | 0.5 |

* One instance not solved within the time limit of 1200 sec        (IBM PowerPC)

in Table 3.3. In Table 3.4 we compare our procedure with the one of Woodruff and Spear-mann [125]. All algorithms are coded in C, and run under UNIX on an IBM PowerPC 601 80 Mhz 41T; instances are generated with the generator described in Section 2.7.

We examine the effectiveness of the rule with two sets of 30 instances with the problem size $(N, J) = \{(3, 22), (5, 18)\}$. In Table 3.2 we display the average ($R_{avg}$) and maximal ($R_{max}$) CPU seconds for B&B[1◇ia-pb] for the different combinations of bounding and dominance rules. Instances are generated with setup-structure *rd* and capacity utilization $\rho$=M. By "+" in the first three columns, we indicate that B&B[1◇ia-pb] uses the occupation bound, considers only regenerative schedules or uses the dominance rule, respectively. Not surprisingly, CPU times are reduced considering only regenerative schedules. The dominance rule again reduces CPU times by an order of magnitude; larger instances cannot be solved without the dominance rule. Simple comparisons are needed to apply the dominance rule and to enumerate only regenerative schedules, with the only drawback that the dominance rule requires a lot of memory. The occupation bound, however, solves one relaxation at each node, and there is a tradeoff between the size of the enumeration tree and the time spent at each node. The last two lines in Table 3.2 show that applying the occupation bound together with the dominance rule may lead to higher CPU times and the best combination of rules is more difficult to find.

In order to determine the maximum problem size $(N, J)$ which is solvable to optimality by B&B[1◇*ia-pb*], we generate instances which are supposed to have no special structure: in Figure 3.4 instances are generated with a random setup matrix (*rd*), a high setup significance $\theta = 1$, and one job per batch. We choose a tight deadline factor as for the capacity utilization $\rho =$H. in this way the instances have no special setup structure, and the number of jobs equals the number of generated batches. The generated schedule contains much setup time and therefore the "real" capacity utilization is supposed to be medium. In the $(N, J)$ diagram in Figure 3.4 a point • (○) indicates, that 30 instances of the respective problem size are solved within 1200 sec and 15 MB memory by B&B[1◇*ia-pb*] without (with) the occupation bound. For a problem size $(N, J + 10)$, however, at least one instance is not solvable within the time and memory limits.

Not surprisingly, instances with a larger number of jobs $J$ can be solved for a smaller number of families $N$. For $(N, J) = (8, 40)$ the occupation bound prevents B&B[1◇*ia-pb*] from exceeding the memory limit but for $(N, J) = (6, 50)$ CPU times *with* the occupation bound are longer so that the time limit is only respected *without* the occupation bound. For the other problem sizes there is no dominant effect of the occupation bound. The results in Figure 3.4 show that we manage to solve rather large instances to optimality, and hence, can provide a solid basis of benchmark instances.

In Table 3.3, the influence of capacity utilization $\rho$ and the setup structure *st-struc* is summarized for instances with $N = 6$ and $J = 36$ on average: they capture well the characteristics of family scheduling problems and the CPU times are moderate. The setup structure *st-struc* is either a group (*gp*), random (*rd*) or sequence (*sq*) structure, we have a setup significance $\theta \approx 0.7$, and batches have a number of jobs out of $DU(1, 2)$. For each $\rho$ - *st-struc* combination we generate 30 instances and compare the [1◇*ia-pb*]-algorithms B&B, DP and C&I. For the exact algorithms B&B[1◇*ia-pb*] and DP[1◇*ia-pb*] average $(R_{avg})$ and maximal $(R_{max})$ computation times are given. We tested two different versions of B&B[1◇*ia-pb*]: with and without the occupation bound (OccBnd).[7]

---

[7]B&B[1◇*ia-pb*] can also be used as a heuristic if enumeration is stopped after a certain time limit,

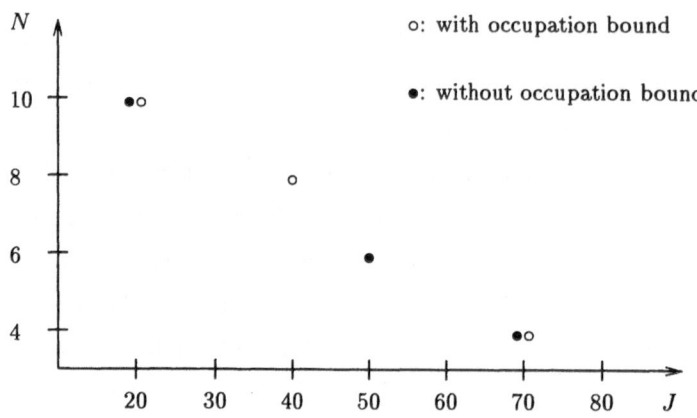

**Figure 3.4:** *Problem Sizes $(N, J)$ solved by B&B[1◇ia-pb] (1200 sec, 15 MB)*

**Table 3.3:** *Influence of st-struc and $\rho$*

| $(N, J) = (6, 36)$ | | CPU time | | | | | Dev. in % and $\#_{inf}$ | | |
|---|---|---|---|---|---|---|---|---|---|
| | | B&B[1◇ia-pb] | | DP[1◇ia-pb] | | C&I[1◇ia-pb] | | | |
| | | OccBnd | | | | | | | |
| $\rho$ | st-struc | $R_{avg}$ | $R_{max}$ | $R_{avg}$ | $R_{max}$ | $R_{avg}$ | $R_{max}$ | $A_{avg}$ | $A_{max}$ | $\#_{inf}$ |
| | gp | 21.1 | 63.1 | 7.2 | 16.5 | 5.2 | 14.4 | 8.4 | 25.4 | 15 |
| H | rd | 15.0 | 34.1 | 8.7 | 18.6 | 21.3 | 82.8 | 9.0 | 17.5 | 13 |
| | sq | 9.4 | 30.1 | 9.3 | 19.0 | 46.5 | 145.4 | 9.0 | 23.4 | 5 |
| | gp | 15.3 | 30.92 | 34.4 | 68.6 | 36.0 | 105.2 | 10.9 | 34.3 | 0 |
| M | rd | 12.3 | 43.32 | 51.6 | 235.2 | 75.2 | 501.1 | 14.2 | 30.6 | 0 |
| | sq | 14.9 | 40.1 | 34.1 | 74.1 | 258.1 | 768.0 | 13.5 | 46.0 | 0 |
| | gp | 15.9 | 38.9 | 84.5 | 202.1 | 162.2 | 525.8 | 10.7 | 33.1 | 0 |
| L | rd | 10.4 | 20.0 | 101.2 | 279.4 | 163.9 | 728.9 | 6.4 | 25.7 | 0 |
| | sq | 11.0 | 27.7 | 50.3 | 119.2 | 360.4 | 928.6 | 8.4 | 20.7 | 0 |

(IBM PowerPC)

For C&I[1◇*ia-pb*], we do not report the (negligible) computation times but the average ($A_{avg}$) and maximal ($A_{max}$) deviation from the optimal objective in %. All instances in Table 3.3 have at least one feasible solution, and with $\#_{inf}$ we denote the number of times (out of 30), the C&I heuristic fails to find one.

For both exact algorithms, computation times increase if $\rho$ decreases: for low capacity utilization the number of sequences to be examined increases, and consequently, both the B&B and the DP algorithm require more time. An exception holds for B&B[1◇*ia-pb*] with the occupation bound: the lower the capacity utilization, the more effective is the occupation bound so that CPU times are in the same order of magnitude for different $\rho$. On the other hand, for $\rho$=H the application of the occupation bound leads to higher CPU times. The heuristic C&I fails to solve instances for $\rho$=H, but always finds a feasible solution for $\rho$=M,L. However, the solution quality of C&I[1◇*ia-pb*] with a maximal deviation of 46 % is not satisfactory.

CPU time is also affected by *st-struc*. For *gp* and $\rho$=H, DP[1◇*ia-pb*] exhibits a better performance than B&B[1◇*ia-pb*], but otherwise the B&B algorithm is more effective. For *sq*, the B&B branching scheme is well suited: if the family $i^s$ is changed from stage $s$ to $s+1$, we first branch to family $i^s+1$. A setup $g \rightarrow i$ with $g > i$ is small in *sq*, we thus first branch to small setups. This explains the small CPU times of B&B[1◇*ia-pb*] for *sq* (and the large ones of DP[1◇*ia-pb*], where we cannot employ a branching rule). However, this effect is not pronounced if the occupation bound is applied. Overall, B&B[1◇*ia-pb*] is more effective than DP[1◇*ia-pb*].

In Table 3.4 we compare our algorithms with a tabu search procedure of Woodruff and Spearman [125], denoted as TSWS. We re-implemented the instance generator of [125] which generated instances for $N = 5$ and $J = 30$; the sequence dependent setup time matrix is given in [125], we denote the setup structure *st-struc* as *woodruff*. Similar to our generator, an initial schedule is generated where the last job completes at time $T$. However, deadlines are assigned in the interval between the completion time of the job

---

which is called *truncated branch and bound* (TBB). But the solution quality of TBB is disappointing, which suggests that a lot of time in B&B[1◇*ia-pb*] is also needed to *find* the optimal solution and not only to prove its optimality.

**Table 3.4:** *Comparison with TSWS of Woodruff and Spearman*

| $(N, J) = (5, 30)$ | | CPU time | | | | Dev. in % and $\#_{inf}$ | | |
|---|---|---|---|---|---|---|---|---|
| | TSWS | B&B[1$\diamond$ia-pb] | | DP[1$\diamond$ia-pb] | | C&I[1$\diamond$ia-pb] | | |
| *st-struc* | $0.1 \cdot R_{avg}$ | $R_{avg}$ | $R_{max}$ | $R_{avg}$ | $R_{max}$ | $A_{avg}$ | $A_{max}$ | $\#_{inf}$ |
| *woodruff* | 100.1 | 0.9 | 6.7 | 2.4 | 16.5 | 15.5 | 94.9 | 20 |

(IBM PowerPC)

and $T$; hence the instances are very constrained and the set of feasible solutions is very small. We generated 30 instances[8] and again report average ($R_{avg}$) and maximal ($R_{max}$) computation times as well as the number of problems $\#_{inf}$ not solved by C&I[1$\diamond$ia-pb]. For TSWS, the values are taken from [125], the code for TSWS was not available.[9]

In [125] a portion of jobs may be left unscheduled ("filler" jobs). The objective of TSWS is to schedule as many filler jobs as possible, such that TSWS and the [1$\diamond$ia-pb] algorithms B&B, DP and C&I, when they find a *feasible* solution, i.e. all jobs are scheduled, actually solve the problem to optimality.[10] For TSWS rather large computation times are reported to find solutions with only a few jobs left unscheduled. For the majority of instances, also C&I[1$\diamond$ia-pb] does not find feasible solutions. The enumerative algorithms, however, solve the instances quite quickly. Though we do not solve the same instances with TSWS as with B&B[1$\diamond$ia-pb] and DP[1$\diamond$ia-pb], the large difference in the CPU times shows that the latter algorithms are more effective; they take advantage of the small solution space. The results suggest that local search based on the neighborhood of job sequences is not well suited for this type of (very constrained) problems since for a long time the search is performed over infeasible solutions.

---

[8]The instance generator in [125] does not guarantee $n_i > 0\ \forall i = 1, \ldots, N$. Therefore, we generated a larger number of instances and took only those with $n_i > 0\ \forall i = 1, \ldots, N$.

[9]CPU times for TSWS are reported on a Dell510 PC, which we assume to be 10 times slower than ours, so $R_{avg}$ for TSWS is multiplied by 0.1 to adjust for this comparison.

[10]Nevertheless, running times in Table 3.4 give the time for the minimization of setup costs. Setup costs are set equal to setup times so that the [1$\diamond$ia-pb] algorithms B&B, DP and C&I minimize total setup time.

## 3.3   Item Availability – Nonpreemptive Batching

In this section we consider problem $[1/fam, st_{g,i}, ia\text{-}npb/\sum w_{(i,j)}C_{(i,j)} + \sum sc_{g,i}]$, written as $[1 \diamond ia\text{-}npb]$ and develop a B&B and a genetic algorithm (GA) for it. The GA, however, only solves the case of sequence independent setups and is denoted as $\text{GA}[1 \diamond ia\text{-}npb, st_i]$.

Unlike the batching type *ia-pb*, in *ia-npb* a batch must not contain idle time. We concentrate on the case where $h_i > 0$; otherwise, there is no difference to the *ia-pb* case, cf. Theorem 2.4. For *ia-npb* the semiactive schedule is not necessarily optimal, which complicates B&B$[1 \diamond ia\text{-}npb]$. However, the sometimes technical analysis is motivated by the fact that B&B$[1 \diamond ia\text{-}npb]$ also solves a well-known lotsizing model, cf. Chapter 4. Another way to solve the problem with sequence independent setups is presented in Section 3.3.2, where $\text{GA}[1 \diamond ia\text{-}npb, st_i]$ solves the BSP in two phases.

### 3.3.1   Branch&Bound

B&B$[1 \diamond ia\text{-}npb]$ uses the enumeration scheme in Section 3.1. We solve a subproblem to derive a schedule from a given sequence in Section 3.3.1.1[11], and the dominance rules in Section 3.3.1.2 are more complicated.

#### 3.3.1.1   Minimum Cost Schedule of a Sequence

For a given sequence $\pi$ we have to decide how to partition $\pi$ into blocks (cf. Definition 2.4), or equivalently, which consecutively *sequenced* jobs should be consecutively *scheduled*. Therefore, in the model formulation in Table 2.5 (p. 26) there is an additional decision variable $P_k$. We have $P_k = 1$ if the job at position $k$ starts a new block, or $P_k = 0$ if it is blocked with the preceding job.

When starting a new block at position $k$ we save earliness costs at the expense of additional setup costs. To derive the minimum cost schedule for *ia-npb*, we start with the semiactive schedule, which is the minimum cost schedule for *ia-pb* (cf. Theorem 2.2) and gives a lower

---

[11]Deriving a schedule from a sequence is called *timetabling* in French [51] (p. 26).

bound for *ia-npb*. More formally: let $\hat{\sigma}$ $(\sigma)$ be the semiactive (minimum cost) schedule for a given sequence $\pi$. Then costs $Z_{BSP}(\hat{\sigma})$ are a lower bound for $Z_{BSP}(\sigma)$.[12]

The following DP recursion forms blocks to find the minimum cost schedule. We define block costs $bc_{k_1,k_2}$ as the cost contribution of a block from position $k_1$ to $k_2$, i.e.

$$bc_{k_1,k_2} = \sum_{k=k_1}^{k_2} w_{(i_{[k]},j_{[k]})}\left(d_{(i_{[k]},j_{[k]})} - C_{(i_{[k]},j_{[k]})}\right) + \sum_{k=k_1+1}^{k_2} sc_{i_{[k-1]},i_{[k]}}.$$

The block size $bs_k$ is the number of jobs blocked with the job $(i_{[k]}, j_{[k]})$ at position $k$.

Let $f_k(b)$ be the cost of a schedule from position $k$ to $J$ if $bs_k = b$. Let $f_k^*$ be the minimum cost schedule and $bs_k^*$ the optimal block size at position $k$. Then, the recurrence equation for determining $f_k^*$ and $bs_k^*$ for $k = J, \ldots, 1$ is:

$$f_k^* = \min_{b=1,\ldots,bs_{k+1}^*+1} \{f_k(b)\} \tag{3.3}$$

$$f_k(bs_k^*) = f_k^*$$

$$f_k(b) = bc_{k,k+b-1} + sc_{0,i_{[k+b]}} + f_{k+b}^*$$

$$f_{J+1}^* = 0; \quad i_{[J+1]} = 0; \quad bs_{J+1}^* = 0 \tag{3.4}$$

In equation (3.3) the state space is reduced based on the following observation: if there is no idle time between position $k$ and $k + 1$ in $\hat{\sigma}$, we schedule job $(i_{[k]}, j_{[k]})$ semiactively; it is then optimal to increment $bs_k^*$, because $sc_{g,i} \leq sc_{0,i}$.

In Table 3.5 we present the computations of equation (3.3) for the example in Figure 2.3 (p. 27), parameters are given in Tables 2.6 and 2.7. For $\pi = ((2,1),(1,1),(3,1),$ $(3,2),(3,3),(2,2),(1,2))$ we derive the minimum cost schedule from the semiactive schedule $\hat{\sigma}$ in Figure 3.5. Schedule $\hat{\sigma}$ contains idle time, and we determine $f_k^*$ and $bs_k^*$ for each position $k$, $k \leq J$ by (3.3).

Consider jobs $(2,2)$, $(3,3)$ and $(3,2)$ at positions 6,5 and 4 in Figure 3.5. All jobs are scheduled semiactively without idle time, and we increment $bs_k^*$, which is denoted by entries (-) in Table 3.5. After job $(3,1)$, $\hat{\sigma}$ has inserted idle time between positions 3 and 4 $(d_{(3,1)} = 10)$ and all different block sizes must be considered to find the minimum

---

[12]The model of *ia-pb* in Table 2.4 is a relaxation of the *ia-npb* model in Table 2.5 where constraints (2.7) and $Y_k$ are omitted. Thus, $Z_{BSP}(\hat{\sigma}) \leq Z_{BSP}(\sigma)$.

**Table 3.5:** *Computations of Equation (3.3) for the Example in Figure 3.5*

| $(i_{[k]}, j_{[k]})$ | $k$ | $f_k^*$ | $bs_k^*$ | $b$ | $f_k(b) =$ | $bc_{k,k+b-1}$ | $+ sc_{0,i_{[k+b]}}$ | $+ f_{k+b}^*$ |
|---|---|---|---|---|---|---|---|---|
| $(0,0)$ | 8 | 0 | 0 | | | | | |
| $(1,2)$ | 7 | 0 | 1 | 1 | $0 =$ | 0 | $+$ 0 | $+$ 0 |
| $(2,2)$ | 6 | - | - | 1 | | - | | |
| | | 1 | 2 | 2 | $1 =$ | 1 | $+$ 0 | $+$ 0 |
| $(3,3)$ | 5 | - | - | 1 | | - | | |
| | | - | - | 2 | | - | | |
| | | 16 | 3 | 3 | $16 =$ | $(5 + 10 + 1)$ | $+$ 0 | $+$ 0 |
| $(3,2)$ | 4 | - | - | 1 | | - | | |
| | | - | - | 2 | | - | | |
| | | - | - | 3 | | - | | |
| | | 18 | 4 | 4 | $18 =$ | $(2 + 5 + 10 + 1)$ | $+$ 0 | $+$ 0 |
| $(3,1)$ | 3 | 23 | 1 | 1 | $23 =$ | 0 | $+$ 5 | $+$ 18 |
| | | | | 2 | $29 =$ | 8 | $+$ 5 | $+$ 16 |
| | | | | 3 | $27 =$ | $(8+8)$ | $+$ 10 | $+$ 1 |
| | | | | 4 | $35 =$ | $(8+8+10+4)$ | $+$ 5 | $+$ 0 |
| | | | | 5 | $36 =$ | $(8+8+10+4+0+6) +$ | 0 | $+$ 0 |
| $(1,1)$ | 2 | - | - | 1 | | - | | |
| | | 28 | 2 | 2 | $28 =$ | 5 | $+$ 5 | $+$ 18 |
| $(2,1)$ | 1 | - | - | 1 | | - | | |
| | | - | - | 2 | | - | | |
| | | 30 | 3 | 3 | $30 =$ | $(2 + 5)$ | $+$ 5 | $+$ 18 |

cost schedule. In Table 3.5 we find $f_3^* = f_3(1)$, i.e. we start a new block after position $k = 3$ and split the group into two batches, as done in $\sigma_{III}$ in Figure 2.3. In order to determine the objective function value, we add $sc_{0,2}$ to $f_1^*$ and obtain a cost of $40[MU]$. Given different setup costs, another minimum cost schedule would be optimal: consider e.g. $st_{0,3} = 10$, $\sigma_{II}$ in Figure 2.3 has lower costs such that $f_3(3)$ would be the minimum

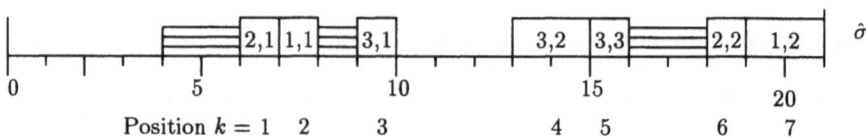

**Figure 3.5:** *Semiactive Schedule*

cost at $k = 3$.

There are two special cases which simplify the general case of equation (3.3). First, if there is no idle time in $\hat{\sigma}$ the blocking problem is trivial: each group in $\hat{\sigma}$ equals one batch, the whole schedule forms a block, and $\hat{\sigma} = \sigma$. Second, if setups are sequence independent, a minimum cost schedule can be derived easier as follows: let the group size $gs_k$ at position $k$ denote the number of consecutively sequenced jobs at positions $r > k$ that belong to the same item as job $(i_{[k]}, j_{[k]})$. Then, for sequence independent setups, equation (3.3) must be evaluated only within $gs_k$. The reasoning is as follows: jobs of *different* groups are leftshifted to be blocked only if we can save setup cost. Then, for consecutive groups of items $g$ and $i$ we would need $sc_{g,i} < sc_{g,0} + sc_{0,i} = sc_{0,i}$, which is not the case for sequence independent setups; therefore, we only need to decide about the leftshift *within* a group.[13]

In the implementation, we compute for each sequence a minimum cost schedule via equation (3.3), and, as for *ia-pb*, we enumerate the sequences. But the computations for equation (3.3) only need to be performed if the semiactive schedule contains idle time behind job $(i^s, j^s)$. In the case of backtracking, the computations have already been done for the $s$-partial schedule to which we backtrack.

### 3.3.1.2 Dominance Rules

Table 3.6 defines further attributes of partial schedules required to describe several dominance rules for *ia-npb*. All jobs which form a block with $(i^s, j^s)$ belong to the set $\mathcal{G}_1(\sigma^s)$,

---

[13]E.g., for sequence independent setups, the computations in Table 3.5 can be restricted to $gs_k$. Then, the maximal block size to be considered at position 3 is 3: the setup between $(3,3)$ and $(2,2)$ would be independent of the idle time between both jobs.

**Table 3.6:** *Attributes of Partial Schedules II*

| | |
|---|---|
| $\mathcal{G}_1(\sigma^s)$ | set of jobs which form the first block of $\sigma^s$ |
| $w_1(\sigma^s)$ | sum of earliness weights of jobs in $\mathcal{G}_1(\sigma^s)$, i.e. |
| | $w_1(\sigma^s) = \sum_{(i,j)\in\mathcal{G}_1(\sigma^s)} w_{(i,j)}$ |

the sum of earliness weights in $\mathcal{G}_1(\sigma^s)$ is denoted as $w_1(\sigma^s)$. For *ia-npb* the difficulty comparing two *s*-partial schedules is that we must take into account the block costs for all extensions of $\sigma^s$ and $\overline{\sigma}^s$. We consider for $\sigma^s$ the maximum, for $\overline{\sigma}^s$ the minimum costs due to blocking; $\sigma^s$ then dominates $\overline{\sigma}^s$ if $c(\sigma^s)$ plus an *upper* bound on block costs is less or equal $c(\overline{\sigma}^s)$ plus a *lower* bound on block costs.

An upper bound on the block costs for $\sigma^s$ is given by $sc_{0,i^s}$ (recall that $sc_{0,i} \geq sc_{g,i}$); then, $\sigma^s$ starts a new block. But a tighter upper bound can be found for start times close to $t(\sigma^s)$: to save costs, we can leftshift all the jobs in $\mathcal{G}_1(\sigma^s)$ (but not more) because after $\mathcal{G}_1(\sigma^s)$ we perform a new setup from the idle machine. $\mathcal{G}_1(\sigma^s)$ is the *largest* block which may be leftshifted. Let $pbt(\sigma^s)$ denote the time where the cost increase due to a leftshift of $\mathcal{G}_1(\sigma^s)$ exceeds $sc_{0,i^s}$. We then have $w_1(\sigma^s)(t(\sigma^s) - pbt(\sigma^s)) = sc_{0,i^s}$ and define the pull-back-time $pbt(\sigma^s)$ of an *s*-partial schedule $\sigma^s$ as follows:

$$pbt(\sigma^s) = t(\sigma^s) - \frac{sc_{0,i^s}}{w_1(\sigma^s)}.$$

Consequently, for time $t$, $pbt(\sigma^s) \leq t \leq t(\sigma^s)$, an upper bound on block costs is given by leftshifting $\mathcal{G}_1(\sigma^s)$; for $t < pbt(\sigma^s)$, block costs are bounded by $sc_{0,i^s}$.

A lower bound on the block costs for $\overline{\sigma}^s$ is given in the same way as for $\sigma^s$, but now we consider the *smallest* block that can be leftshifted, which is simply job $(\overline{i}^s, \overline{j}^s)$.

We can now state the dominance rule for *ia-npb*: we differentiate between the cases $i^s = \overline{i}^s$ (Theorem 3.4) and $i^s \neq \overline{i}^s$ (Theorem 3.5).

**Theorem 3.4** *Consider two s-partial schedules $\sigma^s$ and $\overline{\sigma}^s$ with $\mathcal{AS}^s = \overline{\mathcal{AS}}^s$ and $i^s = \overline{i}^s$. Let $\Delta^i = t(\sigma^s) - t(\overline{\sigma}^s)$ and $\Delta^{ii} = t(\overline{\sigma}^s) - pbt(\sigma^s)$. $\sigma^s$ dominates $\overline{\sigma}^s$ if*

$$(i) \quad t(\overline{\sigma}^s) \leq t(\sigma^s),$$

$$(ii) \qquad c(\overline{\sigma}^s) \geq c(\sigma^s) + \min\{\Delta^i w_1(\sigma^s), sc_{0,i^s}\} \qquad and$$

$$(iii) \qquad c(\overline{\sigma}^s) + \min\{\Delta^{ii} w_{(i^s,j^s)}, sc_{0,i^s}\} \geq c(\sigma^s) + sc_{0,i^s}.$$

**Proof:** As in the proof for *ia-pb* in Theorem 3.3, any completion $\omega^s$ of $\overline{\sigma}^s$ is also a feasible completion of $\sigma^s$ and cannot be scheduled with less costs, due to $(i)$ and $(ii)$. In the following we consider the cost contributions of $\sigma^s$ and $\overline{\sigma}^s$ due to leftshifting when we extend $\sigma^s$ and $\overline{\sigma}^s$:

Consider Figure 3.6 for an illustration of the situation in a time cost diagram. We have $i^s = \overline{i^s}$ and with EDDWF also $(i^s, j^s) = (\overline{i^s}, \overline{j^s})$. The solid line represents the upper bound on block costs for $\sigma^s$. For $pbt(\sigma^s) \leq t \leq t(\sigma^s)$ it is less expensive to leftshift $\mathcal{G}_1(\sigma^s)$, while for $t < pbt(\sigma^s)$ a setup from the idle machine to $i^s$ is performed. The broken line depicts the lower bound on block costs for $\overline{\sigma}^s$. The smallest block that can be leftshifted is the job $(i^s, j^s)$.

In order to claim that $\overline{\sigma}^s$ will never have less costs than $\sigma^s$ due to blocking, we check the costs at points $(ii)$ and $(iii)$: at $(ii)$ we compare the costs at $t(\overline{\sigma}^s)$ while at $(iii)$ we compare the costs at $pbt(\sigma^s)$. Between $(ii)$ and $(iii)$ costs increase linearly, and for $t < pbt(\sigma^s)$ we know that there is monotonous cost increase for $\overline{\sigma}^s$, while costs of $\sigma^s$ no longer increase. Thus, if $(ii)$ and $(iii)$ are fulfilled, cost contributions of $\sigma^s$ in any solution are less than those of $\overline{\sigma}^s$, i.e. there is no completion $\omega^s$ such that $Z_{BSP}(\omega^s, \overline{\sigma}^s) < Z_{BSP}(\omega^s, \sigma^s)$, completing the proof.                                                                 $\square$

For the example in Figure 2.3 (p. 27), Figures 3.7 and 3.8 illustrate the dominance rules. In Figure 3.7, we have $\mathcal{G}_1(\sigma^3) = \mathcal{G}_1(\overline{\sigma}^3) = \{(3,3), (2,2), (1,2)\}$, $t(\sigma^3) \geq t(\overline{\sigma}^3)$ and $pbt(\sigma^3) = 15 - 5/4 = 13.75$. Checking $(ii)$, we have $22 \geq 16 + \min\{1 \cdot 4, 5\}$, while for $(iii)$ we have $22 + \min\{0.25 \cdot 1, 5\} \geq 16 + 5$, so that $\overline{\sigma}^3$ is dominated.

Figure 3.8 illustrates the effect of block costs but with modified data as follows: $sc_{0,3} = 10$ and $d_{3,2} = 15$. Thus $pbt(\sigma^3) = 13 - 10/3$, and $\Delta^{ii} = 10/3$. Checking Theorem 3.4, we have $(ii)$ $16 \geq 15 + \min\{0 \cdot 3, 5\}$, but $(iii)$ $16 + \min\{10/3 \cdot 2, 10\} \geq 15 + 10$ is not fulfilled. Thus, $\sigma^3$ does *not* dominate $\overline{\sigma}^3$, though $c(\sigma^3) < c(\overline{\sigma}^3)$ and $t(\sigma^3) = t(\overline{\sigma}^3)$. Figure 3.8 shows that $c(\sigma^4) > c(\overline{\sigma}^4)$ if $\mathcal{G}_1(\sigma^4)$ is leftshifted. As a consequence, for the application of the dominance rule we must store $w_1(\sigma^s)$ additionally to $t(\sigma^s)$ and $c(\sigma^s)$ for each pair $(\mathcal{AS}^s, i^s)$.

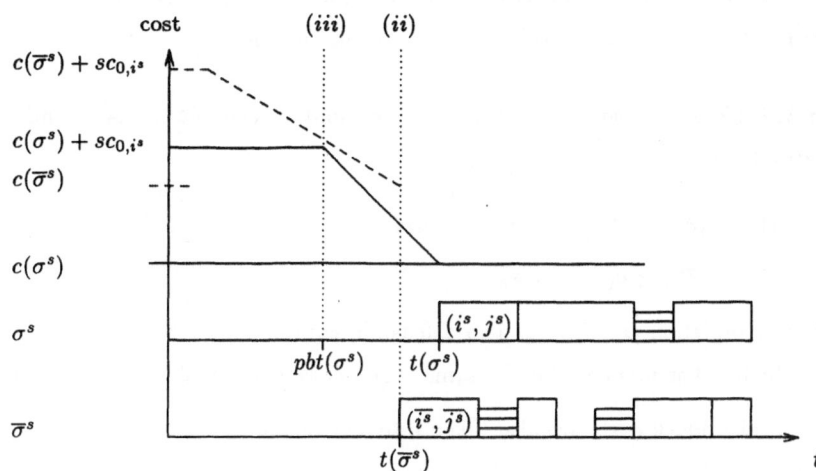

**Figure 3.6:** *Illustration of Theorem 3.4*

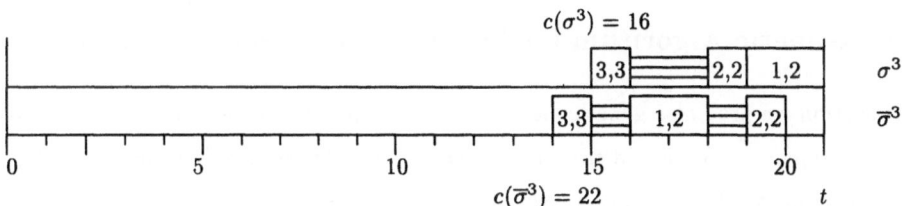

**Figure 3.7:** *Theorem 3.4: $\sigma^3$ dominates $\overline{\sigma}^3$*

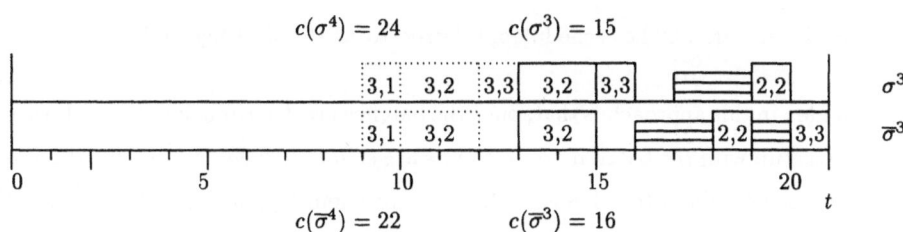

**Figure 3.8:** *Theorem 3.4: $\sigma^3$ does not dominate $\overline{\sigma}^3$*

The second dominance rule for the case $i^s \neq \overline{i^s}$ is similar to Theorem 3.3, but now, instead of $sc_{g,i}$, we must consider $sc_{0,i}$ to take block costs into account.

**Theorem 3.5** *Consider two s-partial schedules $\sigma^s$ and $\overline{\sigma}^s$ with $\mathcal{AS}^s = \overline{\mathcal{AS}}^s$ and $i^s \neq \overline{i^s}$. $\sigma^s$ dominates $\overline{\sigma}^s$ if*

$$
\begin{array}{lll}
(i) & t(\overline{\sigma}^s) + st_{\overline{i^s},i^s} \leq t(\sigma^s) & and \\[2mm]
(ii) & c(\overline{\sigma}^s) \geq c(\sigma^s) + sc_{0,i^s}. &
\end{array}
$$

**Proof:** Any completion $\omega^s$ of $\overline{\sigma}^s$ is also a feasible completion of $\sigma^s$ due to $(i)$, and $\omega^s$ cannot be scheduled at less cost for the same arguments as in the proof of Theorem 3.3.

Now however, also block costs are also taken into account in $(ii)$: an upper bound for the cost contribution of $\sigma^s$, if $\mathcal{G}_1(\sigma^s)$ is leftshifted, is $c(\sigma^s) + sc_{0,i^s}$. A trivial lower bound for the cost contribution of $\overline{\sigma}^s$ is $c(\overline{\sigma}^s)$. Thus $\sigma^s$ dominates $\overline{\sigma}^s$ as any $\omega^s$ completes $\sigma^s$ at lower costs, completing the proof.                                                                    □

## 3.3.2  Genetic Algorithm for Sequence Independent Setups

In this section we present a genetic algorithm[14] for sequence independent setups, denoted as GA[1◇ia-npb,st_i]. We do not use the job sequence as decision variable as in B&B[1◇ia-npb] but decompose the decision in two phases:

- Phase I: how are families partitioned into batches, referred to as Phase I-Batching, and

- Phase II: how are batches scheduled, referred to as Phase II-Scheduling.

The pivotal idea in this approach is that, once the batches are determined in Phase I, we can associate the setup with the batch because setups are sequence independent. Furthermore, this way of decision allows to solve as well the *ba* problem $[1/fam,ba,st_i,d_{(i,j)}/*]$ and the

---

[14]For a general introduction to genetic algorithms cf. e.g. Goldberg [57]. For the design of genetic algorithms for scheduling problems cf. Liepins and Hilliard [85]. In Rubin and Ragatz [107] an application to a problem with sequence dependent setups is presented.

**Table 3.7:** *Decision Variables Phase I-Batching*

| $j_b$ | the $b$-th batch of family $i$ starts with job $(i, j_b)$ |
|---|---|

Resulting batch attributes

| | |
|---|---|
| $p^b_{(i,b)} = st_i + \sum_{l=j_b}^{j_{b+1}-1} p_{(i,l)}$ | batch processing time of the $b$-th batch of family $i$ |
| $d^b_{(i,b)}$ | deadline of the $b$-th batch of family $i$ |

flow-shop problem $[F/fam,ba,st_i,d_{(i,j)}/*]$ with a simple modification. Genetic search is performed for Phase I only, and the problem in Phase II simplifies to a standard scheduling problem with independent jobs. For the batches determined in Phase I, the schedule is determined with a polynomial algorithm in Phase II.

## Phase I-Batching

The decision variables of Phase I are shown in Table 3.7. We assume EDDWF ordering of jobs, thus for Phase I it is sufficient to decide whether a job starts a batch or not to form the batches. The start job $(i, j_b)$ is the first, job $(i, j_{b+1} - 1)$ the last job in the batch $(i, b)$. Then, for each batch $(i, b)$ the attributes batch processing time and batch deadline result from the partitioning of families into batches. The batch processing time $p^b_{(i,b)}$ is simply the sum of processing times of the jobs in the batch plus the (sequence independent) setup time. The batch deadline is the latest time, at which a batch can complete such that all jobs in the batch are available before their deadline. For *ia-npb* the batch deadline is given by equation (2.17), which we repeat here:

$$d^b_{(i,b)} = d_{(i,j_b)} + \sum_{j=j_b+1}^{j_{b+1}-1} p_{(i,j)}.$$

The decisions in Phase I are well suited to be encoded in a genetic string. We choose a binary representation of the Phase I-Batching as the genetic string, cf. Figure 3.9. Each gene represents the batching of one job: if a job is the start job $(i, j_b)$ of a batch the gene is set to one, and to zero otherwise. The jobs $(i, 1), i = 1, ..., N$, are always start jobs and therefore omitted in the genetic string. The string length is $L = \sum_{i=1}^{N} n_i - N$. For the

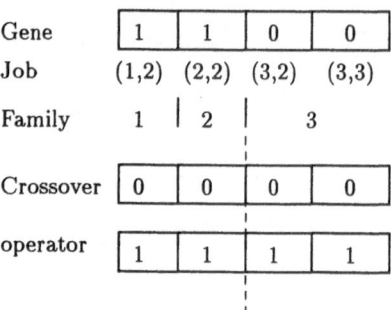

**Figure 3.9:** *Genetic String of Phase I-Batching*

example in Figure 2.3 (p.27), the batches according to the string in Figure 3.9 are provided in Figure 3.10 (recall that setups are sequence independent). We have $L = 2+2+3-3 = 4$, and all $(i,1)$ jobs start a batch. The string determines that jobs $(1,2)$ and $(2,2)$ start a batch (the gene is set to one), and that jobs $(3,2),(3,3)$ are batched with $(3,1)$ (the genes are set to zero). Figure 3.10 displays the batches with the associated setup times at their deadlines. E.g., batch $(i,b) = (3,1)$ has a deadline $d^b_{(3,1)} = 10 + 2 + 1 = 13$ for *ia-npb*. Phase I determines the setup costs of a solution: sequence independent setups costs to family $i$ are incurred for each "1" in the genetic string.

Binary strings are used in a genetic algorithm to search for the best batching decision. Each gene may assume value 0 or 1, independently of the other genes. When we apply the genetic operators crossover or mutation, the new string again describes a new Phase I-Batching decision. Figure 3.9 illustrates the crossover operator: two strings are separated at the same point, and the resulting parts are patched together to a new string. In our case, the string [1100] may be the result of a crossover of the strings [1111] and [0000]; note that all three strings represent valid batching decisions. The mutation operator randomly changes the value of one gene. The encoding allows the development of building blocks and schemes from one generation to the next without having to use "repair"- algorithms. Moreover, we use a small alphabet for the genes and a string which is shorter than the total number of jobs. Encoding the solution in this way is thus well suited to be used in a

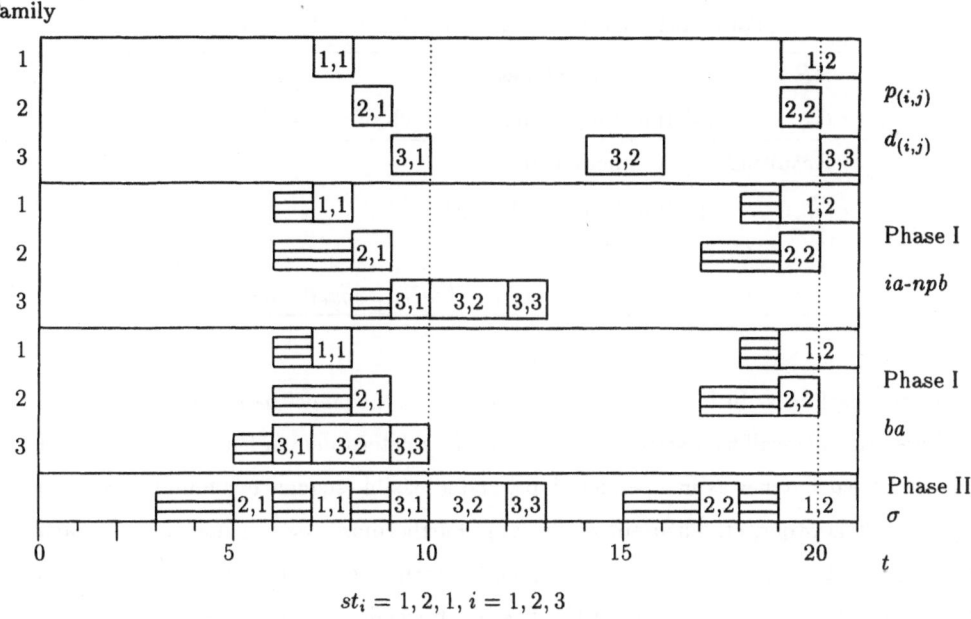

$$st_i = 1, 2, 1, i = 1, 2, 3$$

**Figure 3.10:** *Example for Phase I and Phase II*

genetic algorithm.[15] Note that in a string representing a permutation of jobs a value of a gene must not appear twice, the value of one gene is not "independent" of the value of the other genes. Genetic algorithms based on the encoding of permutations often demonstrate a poor performance, cf. Liepins and Hilliard [85].

## Phase II-Scheduling

For a given batching decision (=genetic string or "individual") Phase II determines a schedule, i.e. the sequence and the completion times of batches and the objective function value (="fitness of the individual") with a deterministic, polynomial algorithm. The decision variables of Phase II are shown in Table 3.8. For a given batch completion time $C^b_{(i,b)}$ the completion times of the jobs in the batch are derived, and consequently,

---

[15]Different concepts for encodings are presented e.g. in Bean [11].

**Table 3.8:** *Decision Variables Phase II-Scheduling*

| | |
|---|---|
| $SB$ | Sequence of all batches |
| $C^b_{(i,b)}$ | completion time of the $b$-th batch of family $i$ |

Resulting job completion times

$C_{(i,j)}$  completion time of job $(i,j)$

$$C_{(i,j_{b+1}-1)} = C^b_{(i,b)},$$

$$C_{(i,l)} = C_{(i,l+1)} - p_{(i,l+1)}, \quad l = j_b, ..., j_{b+1} - 2$$

each job's earliness. In Phase II a feasible batch sequence must be found which minimizes earliness costs (recall that setup costs were already determined in Phase I). In Phase II we can now treat batches as independent jobs, and a feasible sequence is found by sequencing them in EDD order, cf. Theorem 2.9 (p. 47). Furthermore, we can decrease the earliness costs of the EDD schedule of the batches if early batches are scheduled in order of increasing batch weights $w^b_{(i,b)}$, cf. Theorem 2.10. The batch weight is defined in equation (2.18) as

$$w^b_{(i,b)} = \frac{\sum_{l=j_b}^{j_{b+1}-1} w_{(i,l)}}{p^b_{(i,b)}}.$$

The algorithm **Phase II-Scheduling** moves backward in time and minimizes earliness costs heuristically (for similar ideas cf. Ahn and Hyun [5] and Mason and Anderson [90]).

#### Phase II-Scheduling

**Step 1:**  Sort all batches in order of decreasing deadlines. Set the time instant $t$ at the maximal deadline.

**Step 2:**  Select from all unscheduled batches with $d^b_{(i,b)} \geq t$ the one with maximal $w^b_{(i,b)}$.

**Step 3:**  Schedule the batch and set $t$ at the starting time of the batch. If there is no unscheduled batch with $d^b_{(i,b)} \geq t$ set $t$ to the maximal deadline of all unscheduled batches.

**Step 4:**  As long as there are unscheduled batches, `GOTO` **Step 2**, otherwise `STOP`.

We obtain a schedule with the same *occupation* as the EDD schedule, but with lower earliness costs. Consider schedule $\sigma$ in Figure 3.10 for illustration: we schedule the batch for family 3, so time $t = 8$ in **Phase-II Scheduling** is greater or equal the deadline of both batches $(1,1)$ and $(2,1)$. Now **Phase-II Scheduling** selects the next batch to be scheduled not in EDD but in order of increasing batch weights, hence we have $w^b_{(1,1)} = 1/2 > 1/3 = w^b_{(2,1)}$.

In GA$[1\diamond ba, st_i]$ the strings are the individuals in a population which are recombined through mutation and crossover until a prespecified number of generations is reached. Genetic search is performed only for Phase I-Batching. A solution for each string is found with **Phase II-Scheduling**, or, in terms of genetic algorithms, **Phase II-Scheduling** evaluates the string and assigns a fitness to each individual. The selection of individuals for crossover is done randomly, but individuals with a higher fitness have a larger selection probability. We did not conduct numerical experiments to find the best values for the different algorithmic parameters but choose them according to the general recommendations of Goldberg [57]. We only sketch out one implementation detail which speeds up the search for *feasible* solutions: if in a generation no feasible solution is found, the fitness in the next generation is calculated with data where setup costs are decreased with a factor out of $[0,1]$. In this way, we increase the probability that strings with a large number of setups are selected for cross-over; with this strategy, we are able to solve also instances with a small solution space. In comparison, "penalties" for infeasible strings work much less effective to guide the search for feasible solutions.

The general idea to represent the batching decision in a binary string has also been tested in an enumerative algorithm which minimizes the sum of setup costs. In this case, we only need to check if the EDD schedule of batches is feasible in Phase II-Scheduling. The number of batching decisions $2^{J-N}$ is much smaller than the number of EDDWF sequences, but, as we have to perform Phase II-Scheduling at each node, the algorithm was not effective.

## 3.3.3 Computational Results

The problem size which is solvable to optimality by B&B$[1\diamond ia\text{-}npb]$ is much smaller than for *ia-pb*; at each node, we must derive the minimum cost schedule which leads to longer

**Table 3.9:** *Performance of Algorithms for Item Availability - Nonpreemptive Batching*

| $(N, J) = (6, 36)$ | | CPU time B&B[$1\diamond ia$-$npb$] | | Dev. in % and $\#_{inf}$ GA[$1\diamond ia$-$npb$,$st_i$] | | | GA[$1\diamond ia$-$npb$,$st_i$]-short | | |
|---|---|---|---|---|---|---|---|---|---|
| $\rho$ | $\theta$ | $R_{avg}$ | $R_{max}$ | $A_{avg}$ | $A_{max}$ | $\#_{inf}$ | $A_{avg}$ | $A_{max}$ | $\#_{inf}$ |
| H | $h$ | 9.3 | 28.6 | 0.13 | 1.4 | 0 | 0.39 | 3.9 | 3 |
|   | $l$ | 0.4 | 1.1 | 0.00 | 0.0 | 1 | 0.09 | 1.3 | 2 |
| M | $h$ | 143.9 | 382.8 | 0.46 | 3.9 | 0 | 0.83 | 3.6 | 0 |
|   | $l$ | 17.4 | 65.3 | 0.37 | 2.1 | 0 | 0.97 | 3.7 | 0 |
| L | $h$ | 169.1 | 591.0 | 0.45 | 3.4 | 0 | 0.80 | 5.4 | 0 |
|   | $l$ | 34.2 | 108.7 | 0.22 | 2.6 | 0 | 0.20 | 1.5 | 0 |

(IBM PowerPC)

running times. With instances for sequence dependent setups generated in the same way as for Figure 3.4 we managed to solve the problem sizes $(N, J) = (10, 20), (8, 30), (6, 30),$ $(4, 40)$ within 1200 sec and 15 MB computer memory.

For sequence independent setups the performance of B&B[$1\diamond ia$-$npb$] is better, and we can also apply GA[$1\diamond ia$-$npb$,$st_i$]. We generated instances with the generator described in Section 2.7 and varied the setup significance $\theta$. Table 3.9 displays the results. Capacity utilization $\rho$ is high (H), medium (M) and low (L), with low ($l$) and high ($h$) setup significance $\theta$. For each pair $(\rho, \theta)$, we generated 30 instances. Similar to Table 3.3, the influences of $(\rho, \theta)$ are tested for problems with $N = 6$ families and $J = 36$ jobs on average. For B&B[$1\diamond ia$-$npb$] average ($R_{avg}$) and maximal ($R_{max}$) CPU times are given, for GA[$1\diamond ia$-$npb$,$st_i$] we report the average ($A_{avg}$) and maximal ($A_{max}$) deviation from the optimal objective as well as the number of problems not solved (denoted by $\#_{inf}$). The population size is 100 (50) and the number of generations is 200 (100) for GA[$1\diamond ia$-$npb$,$st_i$] (GA[$1\diamond ia$-$npb$,$st_i$]-short). The computation times of the GA directly depend on this parameters and do not exceed 4 (2) CPU seconds.

CPU times for B&B[$1\diamond ia$-$npb$] increase for $\rho =$L, which is the same result as in Table 3.3 for $ia$-$pb$; the number of sequences to be examined increases for lower capacity utilization.

Furthermore, the optimal solutions for *ia-pb* and *ia-npb* do not differ much for $\rho = H$ but may be very different for $\rho = L$. Therefore, for $\rho = L$ the recurrence equation (3.3) is evaluated more often, resulting in increasing CPU times. $\theta$ also strongly affects the CPU times. For $\theta = h$ the feasibility bound is not "sharp" and dominance rules (which must take setups into account) are less effective; consequently, running times are longer.

GA[1◊*ia-npb*,$st_i$] demonstrates a good performance for the instances in Table 3.9. $A_{avg}$ is below one half of a percent for GA[1◊*ia-npb*,$st_i$] (GA[1◊*ia-npb*,$st_i$]-short), and only one instance is not solved by the GA. Due to the selection step and the crossover step, the GA is a non-deterministic algorithm; the instance can be solved if the GA is applied to it several times.[16]

## 3.4 Batch Availability

Batch availability *ba*, finally, is the "most constrained" batching type: all solutions for *ba* are also feasible for *ia-npb* but the inverse is not true. If consecutively sequenced jobs belong to different families $g$ and $i$, we assume for *ba* that there is a setup $g \rightarrow i$ regardless of the idle time between the jobs.

We adapt the B&B algorithm for *ba* and the genetic algorithm for *ba* and sequence independent setups, denoted as B&B[1◊*ba*] and GA[1◊*ba*,$st_i$], respectively. Computational results for both algorithms are presented in Section 3.4.3.

### 3.4.1 Branch&Bound

The enumeration scheme of B&B[1◊*ba*] is described in Section 3.1. In the same way as for *ia-pb*, we must derive a schedule from a given sequence for B&B[1◊*ba*]. For *ba* we derive a *late* and not the minimum cost schedule in Section 3.4.1.1; the adapted dominance rule is presented in Section 3.4.1.2.

---

[16]This is also the reason why $A_{avg}$ and $A_{max}$ for GA[1◊*ia-npb*,$st_i$]-short is smaller than for GA[1◊*ia-npb*,$st_i$] in $(\rho, \theta) = (L, l)$. By chance, GA[1◊*ia-npb*,$st_i$]-short generated a better solution though running time is shorter than for GA[1◊*ia-npb*,$st_i$].

### 3.4.1.1  Late Schedule of a Sequence

A late schedule for a given sequence $\pi$ is the minimum makespan schedule for the mirror problem, cf. the Definition 2.10 (p. 46). We consider only late schedules since among the *feasible* schedules at least one is late (and the feasibility problem is already NP-hard, cf. Section 2.5). But, depending on the cost parameters, a *late* schedule is not necessarily a *minimum cost* schedule. The following DP algorithm derives a late schedule from a given sequence. If there are several late schedules for one sequence, ties are broken with minimum costs in B&B[$1\diamond ba$]. We need the following definitions:

The batch length $btl_{k_1,k_2}$ is the sum of processing times of the jobs from position $k_1$ to $k_2$, i.e.

$$btl_{k_1,k_2} = \sum_{k=k_1}^{k_2} p_{(i_{[k]},j_{[k]})}.$$

The group size $gs_k$ is the number of jobs sequenced after $(i_{[k]},j_{[k]})$ with $i_{[k]} = i_{[k+gs_k-1]}$ and $i_{[k]} \neq i_{[k+gs_k]}$. We redefine the setup time matrix $\tilde{st}_{g,i}$ for batch availability as

$$\tilde{st}_{g,i} = \begin{cases} st_{g,i} & , \quad \text{for } g \neq i \\ st_{0,i} & , \quad \text{for } g = i. \end{cases}$$

At each position $k$ we have to determine the batch size $bts_k^*$ leading to the late schedule. The recurrence equation (3.5) is evaluated within the group size $gs_k > 0$ as for *ia-npb* with sequence independent setups, because setups between *different* families do not depend on the amount of idle time.

Let $ls_k(b)$ be the latest start time of a schedule from position $k$ to $J$ if job $(i_{[k]},j_{[k]})$ starts a batch of size $b$. Time $ls_k^*$ is the maximum of the latest start times among all batch sizes $b$, and $bts_k^*$ the corresponding batch size at position $k$. We evaluate the recurrence equation (3.5) for the determination of $ls_k^*$ and $bts_k^*$ from $k = J, \ldots, 1$.

$$ls_k^* = \max_{b=1,\ldots,gs_k} \{ls_k(b)\} \qquad k = J,\ldots,1 \tag{3.5}$$

$$ls_k(bts_k^*) = ls_k^*$$

$$ls_k(b) = \min\{d_{(i_{[k]},j_{[k]})}, ls_{k+b}^* - \tilde{st}_{i_{[k]},i_{[k+b]}}\} - btl_{k,k+b-1}$$

$$ls_{J+1}^* = \infty; \ i_{[J+1]} = 0; \ gs_{J+1} = 0 \tag{3.6}$$

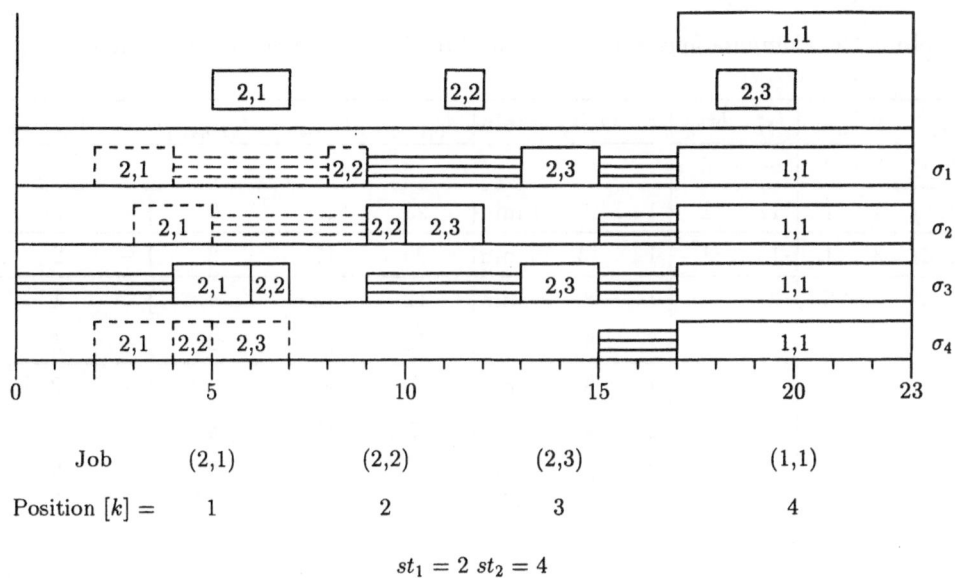

**Figure 3.11:** *Gantt Chart for deriving a Late Schedule*

In equation (3.5) we evaluate $ls_k(b)$ for all batch sizes $b = 1, \ldots, gs_k$ and calculate $ls_k^*$. If different batch sizes $b$ have the same $ls_k(b)$, ties are broken with minimum costs. Batch size $bts_k^*$ determines the $ls_k^*$ which is denoted in the next equation. Calculating $ls_k(b)$ we take the minimum of the batch deadline for $ba$ $d_{(i_{[k]}, j_{[k]})}$, and $ls_{k+b}^* - \tilde{st}_{i_{[k]}, i_{[k+b]}}$ (setup from the last job in the batch to the late schedule at position $k + b$). The recurrence equation is initialized in (3.6).

In Figure 3.11 an example for deriving a late schedule is given, the computations of the recurrence equation (3.5) are shown in Table 3.10. Non-late parts of schedules are depicted in broken lines. We start with $J = 4$ and the initializations given in (3.6). For $k = 1$ we determine the late schedule where job $(2, 1)$ completes at the latest possible time but is still *available* at time 7. E.g., $ls_1(1) = 3$, $(2, 1)$ forms one batch as in $\sigma_2$. But $ls_1(2) = 4 = ls_2^*$ if $(2, 2)$ is batched with $(2, 1)$, $(2, 1)$ completes later and $\sigma_3$ is the late (and the only feasible) schedule for this sequence. For $b = 3$ we have $ls_1(3) = 2$, now $(2, 1)$ starts at time 2 because

**Table 3.10:** *Computations of Equation (3.5) to Derive Late Schedules in Figure 3.11*

| $(i,j)$ | $k$ | $gs_k$ | $ls_k^*$ | $bts_k^*$ | $b$ | $ls_k(b) = \min\{ \ d_{(i_{[k]},j_{[k]})} \ , \ ls_{k+b}^* - \tilde{st}_{i_{[k]},i_{[k+b]}} \ \} - btl_{k,k+b-1}$ | | | | | |
|---|---|---|---|---|---|---|---|---|---|---|---|
| $(0,0)$ | 5 | 0 | $\infty$ | 0 | | | | | | | |
| $(1,1)$ | 4 | 1 | 17 | 1 | 1 | 17 $= \min\{$ | 23 | , | $\infty$ $-$ | 0 | $\} -$ | 6 |
| $(2,3)$ | 3 | 1 | 13 | 1 | 1 | 13 $= \min\{$ | 20 | , | 17 $-$ | 2 | $\} -$ | 2 |
| $(2,2)$ | 2 | 2 | | | 1 | 8 $= \min\{$ | 12 | , | 13 $-$ | 4 | $\} -$ | 1 |
| | | | 9 | 2 | 2 | 9 $= \min\{$ | 12 | , | 17 $-$ | 2 | $\} -$ | 3 |
| $(2,1)$ | 1 | 3 | | | 1 | 3 $\doteq \min\{$ | 7 | , | 9 $-$ | 4 | $\} -$ | 2 |
| | | | 4 | 2 | 2 | 4 $= \min\{$ | 7 | , | 13 $-$ | 4 | $\} -$ | 3 |
| | | | | | 3 | 2 $= \min\{$ | 7 | , | 17 $-$ | 2 | $\} -$ | 5 |

all jobs of family 2 must complete before $d_{(2,1)} = 7$. Figure 3.11 demonstrates that for *ba* batching and sequencing decisions are strongly interdependent: if $\sigma^s$ is a late schedule $\sigma^{s-1}$ is *not* necessarily a late schedule, too.[17]

After the deriving a late schedule, we know $bts_k^*$ for each position $k = J - s + 1$, or equivalently, for each $s$-partial schedule $\sigma^s$ the batch size $bts^s = bts_k^*$ of the late schedule $\sigma^s$. $bts^s$ is needed to formulate the dominance rule for B&B[1◇*ba*].

### 3.4.1.2  Dominance Rule

Again, the dominance rule compares two $s$-partial schedules $\sigma^s$ and $\overline{\sigma}^s$. As for *ia-npb* we differentiate between the cases $i^s = \overline{i^s}$ and $i^s \neq \overline{i^s}$. For $i^s \neq \overline{i^s}$, the *ba* dominance rule equals the one for *ia-npb*, cf. Theorem 3.5. For $\overline{i^s} = i^s$, the dominance rule is weaker than for *ia-npb*: $\sigma^s$ dominates $\overline{\sigma}^s$ if the batch size $bts^s \leq \overline{bts^s}$.

---

[17]More precisely, consider $\sigma_3$ as the late schedule in Figure 3.11. In $\sigma_3$ we have $C_{(2,2)} = 7$ but considering only a 3-partial schedule we have $C_{(2,2)} = 10$ in $\sigma_2$.

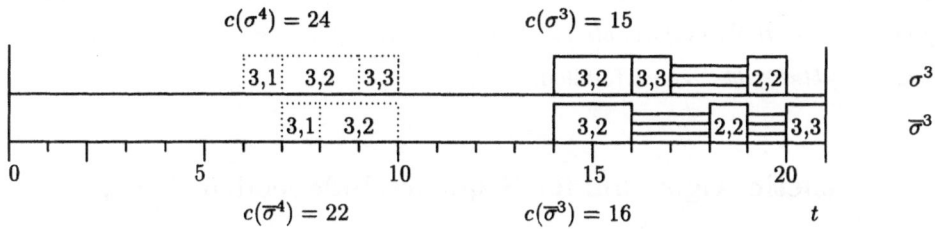

**Figure 3.12:** Theorem 3.6 – $\sigma^3$ does not dominate $\overline{\sigma}^3$

**Theorem 3.6** *Consider two s-partial schedules $\sigma^s$ and $\overline{\sigma}^s$ with $\mathcal{AS}^s = \overline{\mathcal{AS}}^s$ and $\overline{i^s} = i^s$. $\sigma^s$ dominates $\overline{\sigma}^s$ if*

$$(i) \qquad t(\overline{\sigma}^s) \leq t(\sigma^s),$$

$$(ii) \qquad c(\overline{\sigma}^s) \geq c(\sigma^s) \qquad and$$

$$(iii) \qquad \overline{bts^s} \geq bts^s.$$

**Proof:** Let $i^\omega$ be the family of the (last) job in a completion $\omega^s$ of $\overline{\sigma}^s$. If $i^\omega \neq \overline{i^s} = i^s$, then $\sigma^s$ dominates $\overline{\sigma}^s$ for the same arguments as in the proof of Theorem 3.3 due to $(i)$ and $(ii)$.

If $i^\omega = \overline{i^s} = i^s$, we extend $\sigma^s$ with $(i^{s+1}, j^{s+1})$, $i^{s+1} = i^s$. If $(i^{s+1}, j^{s+1})$ is batched with $(i^s, j^s)$ we have again $bts^{s+1} = bts^s + 1 \leq \overline{bts^{s+1}} = \overline{bts^s} + 1$ due to $(iii)$.

Batching $(i^{s+1}, j^{s+1})$ with $(i^s, j^s)$, we must also leftshift all jobs in $\sigma^s$ and $\overline{\sigma}^s$ which are batched with $(i^s, j^s)$, and additional earliness costs are incurred. But due to $(iii)$ we leftshift less jobs in $\overline{\sigma}^s$ than in $\sigma^s$, and additional earliness costs for $\overline{\sigma}^s$ are less than those of $\sigma^s$.

Thus, due to $(ii)$, any $\omega^s$ completes $\sigma^s$ at lower costs so that $\sigma^s$ dominates $\overline{\sigma}^s$, completing the proof. $\square$

If we extend $\sigma^s$ with a job from the same family, we have to take into account that the batch size increases. For $ba$ we store in addition the batch size $bts^s$ of the late schedule $\sigma^s$ for each pair $(\mathcal{AS}^s, i^s)$.

Figure 3.12 illustrates Theorem 3.6. We have $\overline{i^3} = i^3 = 3$. Extending $\sigma^3$ with $(3,1)$, we may get $t(\sigma^4) < t(\overline{\sigma^4})$ because $bts^3 = 2 > \overline{bts^3} = 1$.[18] Thus, $\sigma^3$ must not dominate $\overline{\sigma}^3$ because $(iii)$ $\overline{bts^3} \geq bts^3$ is not fulfilled.

### 3.4.2  Genetic Algorithm for Sequence Independent Setups

GA[1$\diamond$ba,$st_i$] differs from GA[1$\diamond$ia-npb,$st_i$] in Section 3.3.2 only in the way batch deadlines are assigned in Phase I–Batching. For the example in Figure 2.3 (p.27) the batches at their deadlines for $ba$ are displayed in Figure 3.10. We have now $d^b_{(i,b)} = d_{(i,j_b)}$ as in equation (2.16).

Referring to the Gantt–charts of late schedules in Figure 3.11, batches and batch deadlines corresponding to different genetic strings are given in Table 3.11. The schedules $\sigma_1$ to $\sigma_4$ represent the four different ways in which the jobs can be batched for sequence $\pi = ((2,1),(2,2),(2,3),(1,1))$. The string length is now $L = 3 + 1 - 2$, and Table 3.11 displays for each schedule the string for the GA and the batch deadlines. The genetic strings in GA[1$\diamond$ba,$st_i$] may represent *any* Phase I-Batching, and consequently also batching decisions which lead to non–late schedules; e.g., string [11] leads to the non-late schedule $\sigma_1$. Thus, GA[1$\diamond$ba,$st_i$] also searches over non-late schedules, unlike B&B[1$\diamond$ba], and may find better (non-late) solutions.

### 3.4.3  Computational Results

In Table 3.12, computational results are presented in a similar way as in Section 3.3.3. The instance generator now assigns deadlines such that there is a feasible $ba$ schedule. B&B[1$\diamond$ba] solves only much smaller problems than B&B[1$\diamond$ia-npb] for several reasons: first, equation (3.5) for $ba$ must be evaluated much more often than equation (3.3) for $ia$-$npb$. Second, for $ba$ we cannot restrict ourselves to regenerative schedules, which leads to a larger enumeration tree. Third, the dominance rule in Theorem 3.6 is much weaker for

---

[18]In Theorem 3.6 we do not calculate a late schedule when the bounding rule is applied but assume that the batch is enlarged.

**Table 3.11:** *Genetic Strings for Batch Availability Schedules in Figure 3.11*

| $\pi$ | $(2,1)$ | $(2,2)$ | $(2,3)$ | $(1,1)$ |
|---|---|---|---|---|
| GA-String | | 1 | 1 | $\sigma_1$ |
| $d^{ba}$ | 7 | 12 | 20 | 23 |
| GA-String | | 1 | 0 | $\sigma_2$ |
| $d^{ba}$ | 7 | | 12 | 23 |
| GA-String | | 0 | 1 | $\sigma_3$ |
| $d^{ba}$ | | 7 | 12 | 23 |
| GA-String | | 0 | 0 | $\sigma_4$ |
| $d^{ba}$ | | | 7 | 23 |

$i^s = \overline{i^s}$, but it is exactly *this* rule which curtails the enumeration effectively in B&B[$1\diamond ia$-$npb$]. In Table 3.12 therefore only problems with $N = 6$ families and $J = 24$ jobs on average are solved. Instances have zero setup costs ($\gamma = \sum w_{(i,j)}C_{(i,j)}$) such that late schedules are likely to be optimal and B&B[$1\diamond ia$-$npb$] and GA[$1\diamond ba,st_i$] can be compared.

Results in Table 3.12 are quite similar to the ones in Table 3.9. CPU times increase when $\rho$ becomes low while problems with a high setup significance are more difficult to solve. Again, GA[$1\diamond ba,st_i$] demonstrates a good performance for the instances in Table 3.12, $A_{avg}$ and $A_{max}$ are very small, CPU times are always below a second, and a feasible solution is found for all instances. For one instance in $(\rho, \theta) = $(M,$l$) and (L,$h$) a non-late schedule has a smaller objective than the optimum late schedule, such that GA[$1\diamond ba,st_i$] finds a better solution than the B&B algorithm, which only enumerates late schedules. For instances with nonzero setup costs ($\gamma = *$) almost always non-late schedules have a smaller objective than late ones, hence, GA[$1\diamond ba,st_i$] is very well suited for $\gamma = *$.

### 3.4.4 Genetic Algorithm for the Flow-Shop Case

A simple extension makes the GA for the single-machine case applicable to the two-machine flow-shop case ($\alpha = F$), referred to as GA[$F\diamond ba,st_i$], as well. We just solve the single-

**Table 3.12:** *Performance of Algorithms for Batch Availability*

| $(N, J) = (6, 24)$ | | CPU time B&B[$1\diamond ba$] | | \multicolumn{3}{c}{Dev. in % and $\#_{inf}$} | | | | | |
|---|---|---|---|---|---|---|---|---|---|
| | | | | GA[$1\diamond ba, st_i$] | | | GA[$1\diamond ba, st_i$]-short | | |
| $\rho$ | $\theta$ | $R_{avg}$ | $R_{max}$ | $A_{avg}$ | $A_{max}$ | $\#_{inf}$ | $A_{avg}$ | $A_{max}$ | $\#_{inf}$ |
| H | $h$ | 0.9 | 10.5 | 0.01 | 0.4 | 0 | 0.20 | 6.00 | 0 |
| | $l$ | 0.1 | 0.2 | 0.00 | 0.0 | 0 | 0.00 | 0.00 | 0 |
| M | $h$ | 9.1 | 110.4 | 0.10 | 2.8 | 0 | 0.49 | 8.6 | 0 |
| | $l$ | 3.5 | 31.1 | 0.13 | *3.6 | 0 | 0.06 | *1.6 | 0 |
| L | $h$ | 102.8 | 841.9 | 0.28 | *4.00 | 0 | 0.28 | *4.00 | 0 |
| | $l$ | 115.7 | 594.9 | 0.00 | 0.00 | 0 | 0.00 | 0.00 | 0 |

* *positive* deviation for one instance                                        (IBM PowerPC)

machine case twice; therefore, GA[$F\diamond ba, st_i$] is described in this chapter. A B&B algorithm for the more general multi-level case is described in Chapter 5, but only for *ia-pb* and not for *ba*. Computational results for GA[$F\diamond ba, st_i$] are presented in comparison to a two-level lotsizing procedure in Section 4.3.5 (Table 4.10).

The two-machine flow-shop problem is stated as follows: all jobs of one family pass two machines M1 and M2 (in this order), we assume identical processing times $p_{(i,j)}$ and setup times $st_i$ on both machines. A batch on M1 must be completed before processing on M2 can start. In the two-machine case batches on M2 can again be interpreted as jobs (= demands) which can be batched on M1. The deadline of the batch on M1 is the start of its first job on M2. However, the setup can be anticipated on M2, cf. Figure 3.13. The genetic string determines the batching on both machines. In the genetic string we include additional information on how jobs (=batches) on M2 are batched on M1. We enlarge the string alphabet to the values $\{0, 1, 2\}$ for each gene and define its meaning as follows:

$$
\text{value of the gene} = \begin{cases} 1 & , \quad \text{the job starts a batch on M1 } and \text{ M2} \\ 2 & , \quad \text{the job starts a batch only on M2} \\ 0 & , \quad \text{the job } (i, j) \text{ is batched with job } (i, j-1) \end{cases}
$$

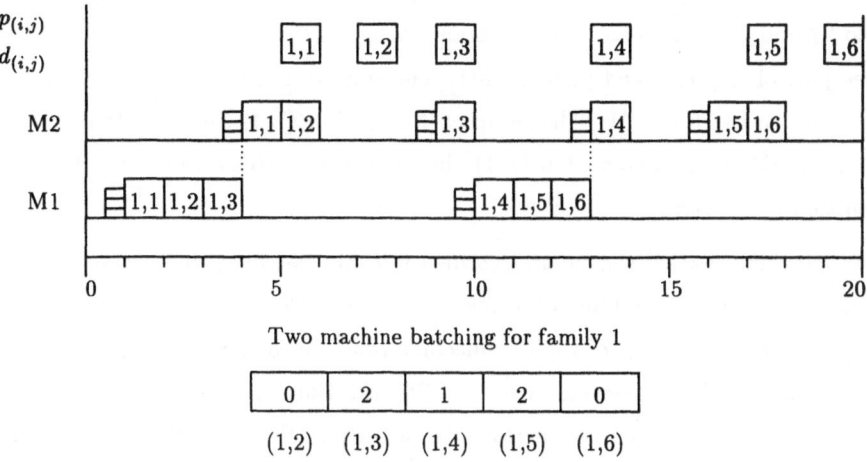

Figure 3.13: *Phase I-Batching for the Two-Machine Flow-Shop Case with ba*

The encoding of the string implicitly considers that those jobs, which are batched on M2, will also be batched on M1. A simple interchange argument in the same way as in Theorem 2.7 (see also the illustration in Figure 2.10 (p. 44)) shows that we do not miss feasibility this way: let $A, B, C$ denote parts of a schedule on M1. Any schedule on M2 defines deadlines for the schedule on M1. Consider a feasible schedule $(C_A, C_{(i,b^1)}, C_B, C_{(i,b^2)}, C_C)$ on M1 where a batch on M2 is split in two batches $(i, b^1)$ and $(i, b^2)$ on M1. The batch on M2 starts after $C_{(i,b^2)}$, thus $(i, b^1)$ and $B$ can be interchanged and schedule $(C_A, \tilde{C}_B, \tilde{C}_{(i,b^1)},$ $C_{(i,b^2)}, C_C)$ is feasible, too. Note, we can leftshift $B$ on M1 because M1 is the first machine. There are no "release dates" on the first machine of the two-machine case. Thus, this reasoning is not valid for the general $M$-machine case.

In Figure 3.13, we illustrate Phase I-Batching (only for family 1) on M1 and M2. The string which determines the batching decisions is given below the Gantt chart. The jobs at their deadline are first batched on M2 in 4 batches. Batches on M2 can again be interpreted as jobs at their deadline for M1. On M1 we have to schedule 4 jobs (=batches on M2) with deadlines 4, 9, 13 and 16. These 4 jobs are batched in 2 batches on M1. Batches on M1 must end before processing on M2 starts: on M2 jobs $(1, 4), (1, 5)$ and $(1, 6)$

form two batches, but one batch on M1. Job $(1,5)$ starts a batch only on M2 (the gene is set to 2), job $(1,4)$ starts a batch on both machines (the gene is set to 1) and batches all three jobs $(1,4), (1,5)$ and $(1,6)$ on M1. The start time 13 on M2 of job $(1,4)$ is the deadline for the batch on M1. The setup on M2 for job $(1,4)$ can be anticipated before the batch on M1 is completed. Job $(1,1)$ always starts a batch on both machines and is thus omitted in the string.

In fact, the string [02120] can be transformed into M1 batching with string [00100] and M2 batching with string [01110] according to the former definition for the single-machine case. The heuristic solution of the two-machine problem in GA$[F \diamond ba, st_i]$ is straightforward: genetic search is performed over the Phase I–Batching strings with an enlarged alphabet. Evaluating a string, we perform the algorithm **Phase-II Scheduling** first for M2 and determine completion times of batches on M2. Then, we can derive deadlines of "jobs" for M1, batch the jobs according to the string and perform **Phase-II Scheduling** on M1.

GA$[F \diamond ba, st_i]$ can also be extended to the case of $M > 2$ machines. Then, the string alphabet is enlarged to $[0; 1; \ldots ; M]$ with a definition corresponding to the one given above.

# Chapter 4

# Discrete Lotsizing and Scheduling by Batch Sequencing

In this chapter we compare the BSP with the discrete lotsizing and scheduling problem (DLSP). We show that BSP and DLSP address the same planning problem, and we examine the question which solution procedures of the different models are better suited to solve it.

In Section 4.1 we introduce the DLSP mathematically and present a review of the relevant literature. The equivalence between both models is motivated, illustrated by means of an example and then formally stated in Section 4.2. Computational results are given in Section 4.3 where the algorithms in Chapter 3 are compared with five procedures reported in literature for the DLSP. Concluding remarks follow in Section 4.4.

## 4.1 Discrete Lotsizing and Scheduling Problem

We present the DLSP with sequence dependent setup times and setup costs (**SDSTSC**).[1] In the DLSP, demand for each item is dynamic and back-logging is not allowed. Before each production run a setup takes place. Setup costs and setup times depend on the sequence of

---

[1]The formulation in Salomon et al. [110] does not exclude undesired schedules and so we use an extended formulation.

**Table 4.1:** *Parameters of the DLSP*

| | |
|---|---|
| $i$ | index of item (=family), $i = 0, \ldots, N$, 0 denotes the idle machine |
| $t$ | index of periods, $t = 1, \ldots, T$ |
| $q_{i,t}$ | demand of item $i$ in period $t$ |
| $h_i$ | holding costs per unit and period of item $i$ |
| $st_{g,i}$ | setup time from item $g$ to item $i$, $g = 0, \ldots, N$; $i = 1, \ldots, N$ |
| $sc^p_{g,i}$ | setup costs per setup period from item $g$ to item $i$, $g = 0, \ldots, N$; $i = 1, \ldots, N$ |
| $sc_{g,i}$ | setup costs from item $g$ to item $i$, $g = 0, \ldots, N$; $i = 1, \ldots, N$ |
| | $sc_{g,i} = sc^p_{g,i} \; \max\{1; st_{g,i}\}$ |

items (sequence dependent). Production serves to meet present or future demand; in the latter case holding costs are incurred. In contrast to the BSP with a continuous time scale, the planning horizon $1, \ldots, T$ is divided into a finite number of (small) periods, indexed by $t$. In each period, at most one item can be produced or a setup is made ("all or nothing production"). An optimal production schedule for the DLSP minimizes the sum of setup and holding costs. As a consequence of the "all or nothing production", setup time and demand are constrained to be multiples of a period length or to the production quantity per period, respectively. In addition to (pure) lotsizing models the DLSP also determines the sequence of lots and generates a schedule as well.[2]

Parameters of the DLSP are presented in Table 4.1, and the decision variables are given in Table 4.2. The DLSP is formally stated in a mixed binary formulation in Table 4.3. Families in the BSP and items in the DLSP are indexed by $i$, holding costs are given by $h_i$. There are sequence dependent setup times $st_{g,i}$ and setup costs $sc_{g,i}$, subject to the assumptions of Remark 2.1. In the DLSP model, we need the setups costs per setup period $sc^p_{g,i}$ which are derived from $st_{g,i}$ and $sc_{g,i}$. The demand matrix $(q_{i,t})$ of the DLSP lists for each period $t$ the exogenous demand for item $i$. Due to the "all or nothing production", we can w.l.o.g. assume $q_{i,t} \in \{0; 1\}$, cf. e.g. Salomon et al. [109] or Magnanti

---

[2]Recall that in the CLSP only the lotsize per period is determined, but not the sequence of lots in a period.

**Table 4.2:** *Decision Variables of the DLSP*

| | |
|---|---|
| $Y_{i,t}$ | 1, if item $i$ is produced in period $t$, and 0 otherwise. $Y_{0,t}$ denotes idle time in period $t$ |
| $V_{g,i,t}$ | 1, if the machine is setup for item $i$ in period $t$, while the previous item was item $g$, and 0 otherwise |
| $I_{i,t}$ | inventory of item $i$ at the end of period $t$ |

and Vachani [89].

In the DLSP, decisions on what is to be done are made in each individual period $t$. We set $Y_{i,t}$ ($V_{g,i,t}$) to one if production (setup from item $g$) takes place for (to) item $i$ in period $t$. The inventory of item $i$ at the end of period $t$ is denoted by $I_{i,t}$. $I_{i,t}$ depends on the decision variable $Y_{i,t}$ and $V_{g,i,t}$, and an equivalent formulation can be given without the inventory variables (cf. e.g. Fleischmann [49]).

In the mixed binary formulation of Table 4.3, the objective (4.1) minimizes the sum of setup costs $sc_{g,i}^p$ (per setup period $st_{g,i}$) and inventory holding costs. Constraints (4.2) express the inventory balance. The "all or nothing production" is enforced by constraints (4.3): in each period, the machine either produces at full unit capacity, undergoes setup for an item, or is idle, i.e. $Y_{0,t} = 1$ for an idle period. For $st_{g,i} = 0$ constraints (4.4) instantiate $V_{g,i,t}$ appropriately. Constraints (4.5) couple setup and production whenever $st_{g,i} > 0$: if item $i$ is produced in period $t$ and item $g$ in period $t - \tau - 1$ then the decision variable $V_{g,i,t-\tau} = 1$ for $\tau = 1, \ldots, st_{g,i}$. Constraints (4.6) enforce the correct length of the string of setup variables $V_{g,i,t-\tau}$ for $st_{g,i} > 1$. However, for $st_{g,i} > 0$ we have to exclude the case $Y_{g,t-1} = Y_{i,t} = 1$ without setting any $V_{g,i,t} = 1$; this is done by constraints (4.7). Constraints (4.8) prevent any back-logging. Finally, the variables $I_{i,\tau}$, $V_{g,i,\tau}$, and $Y_{i;\tau}$ are initialized for $\tau \leq 0$, by constraints (4.11).

Clearly, the model in Table 4.3 contains the DLSP with sequence dependent setups costs but zero setup times (**SDSC**), or sequence independent setup times and setup costs (**SISTSC**) as special cases: either $st_{g,i} = 0$ or $st_{g,i} = st_i$, $g \neq i$, analogously for setup costs. SDSTSC, SDSC and SISTSC cover the case of item availability–nonpreemptive

**Table 4.3:** *Model of the DLSP*

$$\text{Min} \quad Z_{DLSP} = \sum_{i=1}^{N} \sum_{t=1}^{T} \left( \sum_{g=0}^{N} sc_{g,i}^{p} V_{g,i,t} + h_i I_{i,t} \right) \tag{4.1}$$

subject to

$$I_{i,t-1} + Y_{i,t} - q_{i,t} = I_{i,t} \qquad\qquad i = 1, \ldots, N; \quad t = 1, \ldots, T \tag{4.2}$$

$$\sum_{i=0}^{N} Y_{i,t} + \sum_{\{g,i \mid st_{g,i} > 0, g \neq i\}} V_{g,i,t} = 1 \qquad t = 1, \ldots, T \tag{4.3}$$

$$V_{g,i,t} \geq Y_{g,t-1} + Y_{i,t} - 1 \qquad \left\{ \begin{array}{l} g = 0, \ldots, N; \quad i = 1, \ldots, N; \quad g \neq i; \\ st_{g,i} = 0; \quad t = 1, \ldots, T; \end{array} \right. \tag{4.4}$$

$$V_{g,i,t-\tau} \geq Y_{i,t} + Y_{g,t-\tau-1} - 1 \qquad \left\{ \begin{array}{l} g = 0, \ldots, N; \quad i = 1, \ldots, N; \quad g \neq i; \\ st_{g,i} > 0 \quad t = 1, \ldots, T; \\ \tau = 1, \ldots, st_{g,i} \end{array} \right. \tag{4.5}$$

$$V_{g,i,t-\tau} \geq Y_{i,t} + V_{g,i,t-1} - 1 \qquad \left\{ \begin{array}{l} g = 0, \ldots, N; \quad i = 1, \ldots, N; \quad g \neq i; \\ st_{g,i} > 1; \quad t = 1, \ldots, T; \\ \tau = 2, \ldots, st_{g,i} \end{array} \right. \tag{4.6}$$

$$Y_{g,t-1} + Y_{i,t} \leq 1 \qquad \left\{ \begin{array}{l} g = 0, \ldots, N; \quad i = 1, \ldots, N \quad g \neq i; \\ st_{g,i} > 0; \quad t = 1, \ldots, T \end{array} \right. \tag{4.7}$$

$$I_{i,t} \geq 0 \qquad\qquad i = 1, \ldots, N; \quad t = 1, \ldots, T \tag{4.8}$$

$$V_{g,i,t} \in \{0; 1\} \qquad \left\{ \begin{array}{l} g = 0, \ldots, N; \quad i = 1, \ldots, N; \quad g \neq i; \\ t = 1, \ldots, T \end{array} \right. \tag{4.9}$$

$$Y_{i,t} \in \{0; 1\} \qquad\qquad i = 0, \ldots, N; \quad t = 1, \ldots, T \tag{4.10}$$

$$I_{i,\tau} = V_{g,i,\tau} = Y_{i,\tau} = 0 \qquad\qquad g, i = 0, \ldots, N; \quad \tau \leq 0 \tag{4.11}$$

batching: if a batch is preempted, e.g. $Y_{i,t} = Y_{i,t+2} = 1 \neq Y'_{i,t+1} = 0$, a new setup is needed for production in period $t + 2$. A model formulation for the single-level DLSP with batch availability (**BA**) is presented by Brüggemann and Jahnke [19] and extended for the two-level case with $ba$ (**BA2LV**) in Brüggemann and Jahnke [20]. They present formulations which require a lot of additional variables.[3] The simulated annealing approach in [19] and [20] is examined in Section 4.3.4 and 4.3.5.

In the literature, a variant of the DLSP with zero setup times and sequence independent setup costs was first considered by Lasdon and Terjung [80], with an application to production scheduling in a tire company. Glassey [56] considers only changeover costs in his model. The generic DLSP is introduced by Fleischmann [49]. A lower bound is determined by lagrangean relaxation and a heuristic, which uses the lagrangean multipliers, determines an upper bound; the optimal solution is found by a B&B algorithm which uses upper and lower bound. Computational experience shows that both bounds are very tight so that only very few nodes must be enumerated. Magnanti and Vacchani [89] examine a problem which is closely related to the DLSP and propose polyhedral methods. Gascon and Leachman [54] and Leachman et al. [82] propose DP algorithms for the DLSP with item availability–preemptive batching, but they report results only for small problem sizes. Ahmadi et al. [2] consider a parallel machine DLSP where orders are released at different times from an upstream stage, and setup costs and waiting costs are to be minimized. They propose different heuristics to generate near optimal solutions. In this "reverted" DLSP the feasibility problem is polynomially solvable, and penalties are incurred for the quantities which are not produced within the planning horizon.

SISTSC is considered by Cattrysse et al. [24] who present a mathematical programming based approach (cf. also Section 4.3.1).

An earlier work dealing with SDSC is the one of Popp [100]. Fleischmann [50] incorporates ideas from solution procedures for vehicle routing problems to solve the SDSC, cf. also Section 4.3.2. Salomon et al. [110] extend Fleischmann's approach to solve SDSTSC; a more detailed description of their exact algorithm is presented in Section 4.3.3.

---

[3]In the $ba$ case, additional variables for BA and BA2LV are needed to determine ($i$) the batch completion and ($ii$) the batch size (which both depend on $Y_{i,t}$) to state the inventory balance constraints.

The complexity of the DLSP is considered in Salomon et al. [109]. One result is that SISTSC is NP–hard, which is proven with reference to Bruno and Downey [22]. Remarks on the results in [109] are given by Brüggemann and Jahnke [21] and Brüggemann [18]. In Salomon [108] and [109] a classification scheme for the DLSP is developed, where [108] also gives a general overview of the DLSP and lotsizing models with multi-level product structures.

## 4.2   Equivalence of DLSP and BSP

The main observation that motivated us to solve the DLSP as a special case of the BSP is that the $(q_{i,t})$-matrix is sparse, especially if setup times are significant: the maximum number of *positive* entries in $(q_{i,t})$ is $T$ $(q_{i,t} \in \{0;1\})$ while the *total* number is $N \times T$.

The basic idea is to interpret *items* in the DLSP as *families* in the BSP (there is no setup if the same item is scheduled in consecutive periods, i.e. in a *batch*) and to consider the *demands* in the DLSP as *jobs* in the BSP with a deadline and a processing time (cf. also Jordan and Drexl [71]). We derive instances of the BSP from DLSP instances in the following way: setup times and setup costs in the BSP and DLSP are identical, and the job attributes of the BSP instances are derived from the $(q_{i,t})$-matrix by Definitions 4.1 and 4.2. The batching type in the DLSP is *ia-npb*: a new setup is needed if idle time preempts the batch.

**Definition 4.1** *An instance of $[1/fam,ia\text{-}npb,st_{g,i},d_{(i,j)},p_{(i,j)} = 1/\sum w_{(i,j)}C_{(i,j)} + \sum sc_{g,i}]$ derived from a DLSP instance is denoted as BSPUT(DLSP) (unit time jobs). Each positive entry, $q_{i,t} = 1$, is a job with $p_{(i,j)} = 1$, $w_{(i,j)} = h_i$ and $d_{(i,j)} = t$. For each item $i$ there are $n_i = \sum_{t=1}^{T} q_{i,t}$ jobs.*

**Definition 4.2** *An instance of $[1/fam,ia\text{-}npb,st_{g,i},d_{(i,j)}/\sum w_{(i,j)}C_{(i,j)} + \sum sc_{g,i}]$ derived from a DLSP instance is denoted as as BSP(DLSP). A sequence of consecutive "ones" in the $(q_{i,t})$-matrix, i.e. $q_{i,t} = 1$, $t = t_1,\ldots,t_2$ is a job with $p_{(i,j)} = t_2 - t_1 + 1$, $w_{(i,j)} = h_i p_{(i,j)}$, and $d_{(i,j)} = t_2$. The number of times that we have a sequence of consecutive ones for an item $i$ defines $n_i$.*

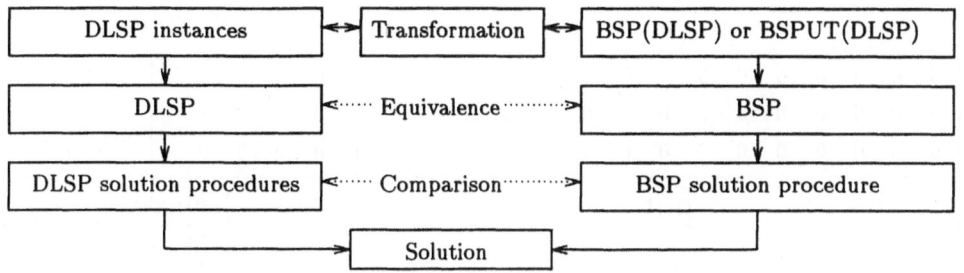

**Figure 4.1:** *Comparison of DLSP and BSP*

Figure 4.1 provides the framework for the comparison: after the transformation of DLSP instances into BSP instances, the performance of solution procedures and the quality of the solutions can be compared since both models are equivalent. The difference between the approaches is as follows: in the DLSP decisions are made anew in each individual period $t$ with decision variables $Y_{i,t}$ and $V_{g,i,t}$ (cf. Table 4.2). In the BSP, we decide how to schedule jobs, i.e. we decide about the completion times of the jobs. BSP and DLSP address the same underlying planning problem, but use different decision variables. Brüggemann and Jahnke [21] and Brüggemann [18] make another observation which concerns the transformation of instances: a DLSP instance may be *not* polynomially bounded in size while the size of the BSP(DLSP) instance is polynomially bounded. On that account, in [21] and [18] it is argued that the $(q_{i,t})$-matrix is not a "reasonable"encoding for a DLSP instance in the sense of Garey and Johnson [52], because BSP(DLSP) describes a problem instance in a shorter way.

In Figure 4.2 we recall the example from Figure 2.3 to illustrate the equivalence between both models. Parameters for the setups and holding costs are the same as in Table 2.6. Figure 4.2 shows the demand matrix $(q_{i,t})$ for DLSP and the jobs at their deadline for BSPUT(DLSP) and BSP(DLSP).

For BSPUT(DLSP) we interpret each entry of "one" as a job $(i,j)$ and derive $d_{(i,j)}$. Processing times $p_{(i,j)}$ equal one for all jobs. We summarize the BSPUT(DLSP) parameters in Table 4.4. An optimal solution for DLSP with $h_2 = 3$ is the string $\nu^a$ in Figure 4.2,

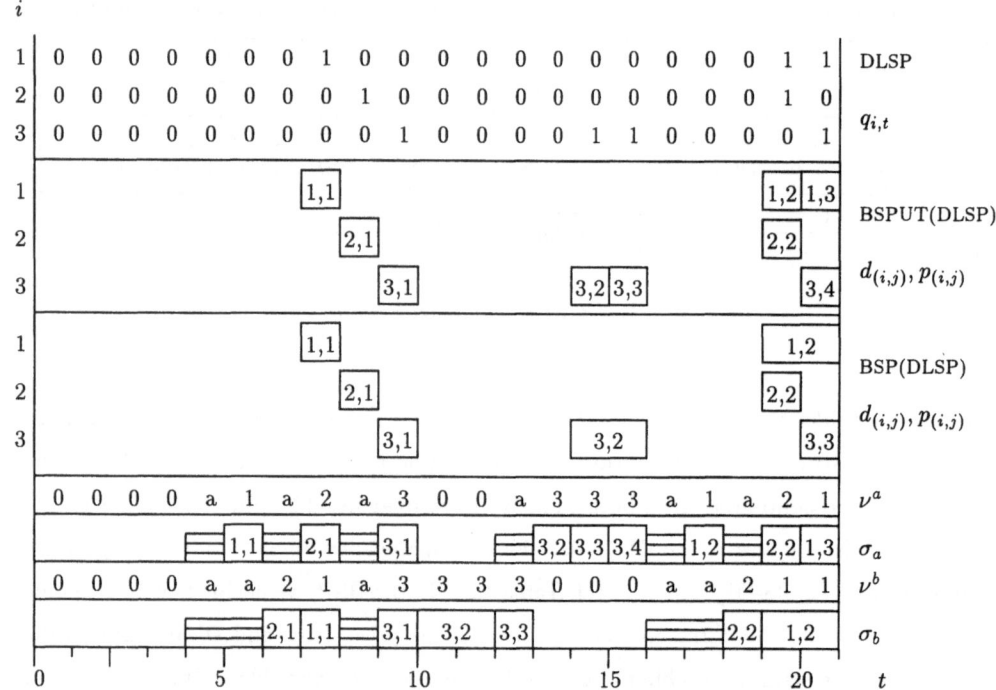

**Figure 4.2:** *DLSP, BSPUT(DLSP) and BSP(DLSP)*

with entries [0,a,1,2,3] for idle or setup time, or for production of the different items, respectively. This solution is represented by $\sigma_a$ for BSPUT(DLSP) and displayed in Table 4.4. Both solutions have an optimal objective function value of 44[MU].

In BSP(DLSP) consecutive "ones" in the demand matrix ($q_{i,t}$) are linked to one job. The number of jobs is thus smaller in BSP(DLSP) than in BSPUT(DLSP). E.g. jobs $(1,2)$ and $(1,3)$ in BSPUT(DLSP) are linked to one job $(1,2)$ in BSP(DLSP). However, a solution of BSP(DLSP) cannot represent, for example, $\sigma_a$ in Figure 4.2, where job $(1,2)$ is split.[4] For different cost parameters $h_i = 1$, $i = 1,2,3$, and $sc_{0,3} = 10$, $\nu^b$ is the optimal solution for the DLSP; now holding costs are the same for all items. The same solution

---

[4]But note that $\sigma_a$ is not a regenerative schedule since job $(1,3)$ starts a batch at $d_{(1,2)} = 20$.

Table 4.4: *BSPUT(DLSP) Instance and Solution*

| $(i,j)$ | $(1,1)$ | $(1,2)$ | $(1,3)$ | $(2,1)$ | $(2,2)$ | $(3,1)$ | $(3,2)$ | $(3,3)$ | $(3,4)$ | |
|---|---|---|---|---|---|---|---|---|---|---|
| $d_{(i,j)}$ | 8 | 20 | 21 | 9 | 20 | 10 | 15 | 16 | 21 | |
| $p_{(i,j)}$ | 1 | 1 | 1 | 1 | 1 | 1 | 1 | 1 | 1 | |
| $w_{(i,j)}$ | 1 | 1 | 1 | 3 | 3 | 1 | 1 | 1 | 1 | |
| $k$ | 1 | 2 | 3 | 4 | 5 | 6 | 7 | 8 | 9 | |
| $(i_{[k]}, j_{[k]})$ | $(1,1)$ | $(2,1)$ | $(3,1)$ | $(3,2)$ | $(3,3)$ | $(3,4)$ | $(1,2)$ | $(2,2)$ | $(1,3)$ | |
| $C_{(i_{[k]},j_{[k]})}$ | 6 | 8 | 10 | 14 | 15 | 16 | 18 | 20 | 21 | $\sigma_a$ |

is the schedule $\sigma_b$ for BSP(DLSP). Again, the optimal objective function value is 44[MU] for $\nu^b$ and $\sigma_b$.

The example shows, that the same solution can be obtained from different models. In the sequel, we formally analyze the **equivalence** between the DLSP and the BSP with *ia-npb*.

**Definition 4.3** *A vector $\nu = (\nu_1, \nu_2, \ldots, \nu_T)$ represents a solution of DLSP as a period-item assignment specifying the decision of what to do in each period, i.e. $\nu_t = i$, if $Y_{i,t} = 1$ or $\nu_t = a$ if $V_{g,i,t} = 1$. $\nu$ is called a solution if $Y_{i,t}$ and $V_{g,i,t}$ are feasible for DLSP.*

*We call a solution $\nu = (\nu_1, \nu_2, \ldots, \nu_T)$ of DLSP and a solution $\sigma$ of BSPUT(DLSP) corresponding solutions if for each point in time $t$*

1. *the family $i$ of the job being processed at $t$ equals the entry in $\nu$, i.e. $\nu_t = i$.*

2. *where a setup is performed in $\sigma$ we have $\nu_t = a$.*

3. *where the machine is idle in $\sigma$ we have $\nu_t = 0$.*

Due to the "all or nothing production" a solution in terms of the decision variables $Y_{i,t}$ and $V_{g,i,t}$ can be represented by a vector $\nu = (\nu_1, \nu_2, \ldots, \nu_T)$ according to Definition 4.3. The DLSP solution $\nu$ and the corresponding BSPUT(DLSP) solution $\sigma$ describe the same decision: in Figure 4.2, $\nu^a$ corresponds to $\sigma_a$ and $\nu^b$ corresponds to $\sigma_b$. By an EDDWF ordering completion times in $\sigma$ are derived from $\nu$ or entries in $\nu$ are derived from $\sigma$.

**Theorem 4.1** $\sigma$ *is a solution for BSPUT(DLSP) if and only if the corresponding solution $\nu$ is a solution of DLSP and $\nu$ and $\sigma$ have the same objective function value.*

**Proof:** We first prove that the constraints of DLSP and BSPUT(DLSP) define the same set of solutions.

In DLSP, constraints (4.2) and (4.8) enforce that the decision variables $Y_{i,t}$ are equal to one between zero and the occurrence of the demand. The sequence on the machine – the sequence dependent setup times taken into account – is expressed by constraints (4.3) to (4.7). In the BSP, this is covered by constraints (2.6) in the model of Table 2.5. We schedule each job between zero and its deadline, represented by constraints (4.2) and (4.8) in DLSP. All jobs are processed on a single-machine taking into account sequence dependent setup times.

Second, we prove that the objective functions (4.1) and (2.5) assign the same value to corresponding solutions $\nu$ and $\sigma$:

The cumulated inventory for an item $i$ (over the planning horizon $1, \ldots, T$) equals the cumulated earliness of family $i$ and job weights equal the holding costs, i.e. $h_i = w_{(i,j)}$ for BSPUT(DLSP). Thus, the terms $\sum w_{(i,j)}(d_{(i,j)} - C_{(i,j)})$ and $\sum h_i I_{i,t}$ are equal for corresponding solutions $\nu$ and $\sigma$.

Some more explanation is necessary to show that corresponding solutions $\nu$ and $\sigma$ have the same setup costs. Consider a setup $g \to i$, $g, i \neq 0$, without idle time in $\sigma$: we then have $sc_{g,i} = sc_{g,i}^p max\{1; st_{g,i}\}$ whereas in $\nu$ we need $st_{g,i}$ consecutive "ones" in $V_{g,i,t}$. On the other hand, for the case of inserted idle time in *ia-npb*, we have a setup from the idle machine (enforced by the decision variable $P_k$ for the BSP) while in $\nu$ there are $st_{0,i}$ consecutive "ones" in $V_{0,i,t}$. Thus, the terms $\sum sc_{i_{[k-1]},i_{[k]}} + P_k(sc_{0,i_{[k]}} - sc_{i_{[k-1]},i_{[k]}})$ and $\sum sc_{g,i}^p V_{g,i,t}$ are equal for corresponding solutions $\nu$ and $\sigma$.

Therefore, corresponding solutions $\nu$ and $\sigma$ incur the same holding and the same setup costs, which proves the theorem.                                              □

As to the batching type batch availability, a formal proof that the constraints define the same set of decisions cannot be given since we do not present a model for *ba*. However, the objective (4.1) for *ba* is the same as for *ia-npb*, such that this part of Theorem 4.1 remains the same for *ba* and *ia-npb*.

**Corollary 4.1** *A schedule $\sigma$ is optimal for BSPUT(DLSP) if and only if the correspond-*
*ing solution $\nu$ is optimal for DLSP.*

**Proof:** Follows from Theorem 4.1.                                               □

Corollary 4.1 is the desired result and states the equivalence between DLSP and BSP
for BSPUT(DLSP) instances. We can thus solve DLSP by solving BSPUT(DLSP). In
general, however, the more attractive option will be to solve BSP(DLSP) because there
the number of jobs is smaller. We first show that we do not lose feasibility considering
BSP(DLSP) instances.

**Corollary 4.2** *If there is a feasible solution $\nu$ of DLSP, there is also a feasible solution $\sigma$*
*of BSP(DLSP).*

**Proof:** As there is a feasible $\nu$, we know that there is a corresponding $\tilde{\sigma}$ for BSPUT(DLSP),
which is feasible as well. By Theorem 2.7 there is a feasible regenerative schedule $\sigma$, which
represents a BSP(DLSP) solution: all unit time jobs $(i,j)$ and $(i,j+1)$ with $d_{(i,j)} + 1 =$
$d_{(i,j+1)}$ are scheduled consecutively in a regenerative schedule, completing the proof.    □

Analogously to Theorem 2.8, regenerative schedules are optimal for identical holding costs.

**Corollary 4.3** *Let the holding costs for all items be identical in DLSP, i.e. $\forall i : h_i = h$.*
*If $\sigma$ is an optimal solution of BSP(DLSP), the corresponding solution $\nu$ is optimal for*
*DLSP.*

**Proof:** The proof proceeds analogous to the one above. By Theorem 2.8 there is an
optimal regenerative schedule $\sigma$ for BSPUT(DLSP); the corresponding $\nu$ is optimal for
DLSP by Corollary 4.1. A regenerative schedule $\sigma$ also represents a BSP(DLSP) solution,
completing the proof.                                                             □

To summarize we have so far obtained the following results:

1. DLSP and BSP are equivalent for BSPUT(DLSP).

2. Feasibility of BSP(DLSP) implies feasibility of DLSP.

3. For identical holding costs, an optimal solution of BSP(DLSP) is optimal for DLSP.

The theoretical difference between BSP(DLSP) and DLSP in our third result has only a small effect when instances with different holding costs are solved. Computational results in Section 4.3.3 show that almost always an optimal regenerative solution of BSPUT(DLSP) exists which is found by solving BSP(DLSP).

## 4.3    Comparison with Procedures to Solve Variants of the DLSP

From the analysis in Section 4.2 we know that we address the same planning problem in BSP and DLSP and that we yield corresponding solutions. Consequently, in this section we compare the performance of algorithms solving the BSP with procedures for solving variants of the DLSP. The comparison is made with the DLSP instances used to test the DLSP procedures; we take the instances provided by the cited authors and solve them as BSP(DLSP) or BSPUT(DLSP) instances, cf. Figure 4.1. An exception is made for reference [50] where we generated instances with the generator of Section 2.7.

The different DLSP variants are summarized in Table 4.5. For the DLSP, in the first column the reference, in the second the DLSP variant is displayed. The fourth column denotes the proposed algorithm, the third column shows whether computational results for the proposed algorithm are reported for identical holding costs or not. Depending on the holding costs, the different DLSP variants are solved as BSP(DLSP) or BSPUT(DLSP) instances, and different algorithms of Chapter 3 are applied. With the exception of reference [110] the DLSP procedures are tested with identical holding costs, such that regenerative schedules are optimal in [24] and [50]. In references [19] and [20] for $ba$, we have the result of Theorem 2.10: in optimal schedules some batches can be ordered according to the batch weights.

### 4.3.1    Sequence Independent Setup Times and Costs (SISTSC)

In Cattrysse et al. [24] a mathematical programming based procedure to solve SISTSC is proposed. Cattrysse et al. refer to their procedure as dual ascent and column generation

**Table 4.5:** *Solving Different DLSP Variants as a BSP*

| DLSP | | | | BSP | |
|------|---------|-----------------|-----------|----------------------|------------------------|
| Author | Variant | Holding costs | Algorithm | Algorithms | Instances / Structural Properties |
| Cattrysse et al. [24] | SISTSC | $h_i = 1$ | DACGP | B&B[1◇$ia$-$npb$] <br> GA[1◇$ia$-$npb$,$st_i$] | BSP(DLSP) / regenerative, Theorem 2.8 |
| Fleisch-mann [50] | SDSC | $h_i = 1$ | TSPOROPT | B&B[1◇$ia$-$npb$] | BSP(DLSP) / regenerative, Theorem 2.8 |
| Salomon et al. [110] | SDSTSC | $h_i > 0$ | TSPTWA | B&B[1◇$ia$-$npb$] | BSPUT(DLSP) / EDDWF, Theorem 2.6 |
| | | $h_i = 0$ | TSPTWA | B&B[1◇$ia$-$npb$] | BSP(DLSP) / block, Theorem 2.4 |
| Brüggemann and Jahnke [19] | BA | $h_i = 1$ | BRJSA | B&B[1◇$ba$] <br> GA[1◇$ba$,$st_i$] | BSP(DLSP) / Theorem 2.10 |
| Brüggemann and Jahnke [20] | BA2LV | $h_i = 1$ | BRJSA2LV | GA[$F$◇$ba$,$st_i$] | BSP(DLSP) / Theorem 2.10 |

procedure, denoted as DACGP. The DLSP is first formulated as a set partitioning prob-lem (SPP), where the columns represent the production schedule for one item $i$; the costs of each column can be calculated separately because setups are sequence independent. DACGP then computes a *lower bound* for the SPP by column generation; new columns can be generated from solving a single item subproblem by a (polynomial) DP recursion.

In DACGP a feasible solution, i.e. an *upper bound*, may be found during column generation whenever a solution is found where the production schedules contain no slack[5].
Furthermore, an upper bound is calculated with the enumerative algorithm of Garfinkel
and Nemhauser [53] from the columns generated so far. If in both cases no feasible solution
is found, an attempt is made with a simplex based procedure[6].

The (heuristic) DACGP generates an upper and a lower bound whereas B&B[1⋄*ia-npb*]
solves BSP(DLSP) to optimality. GA[1⋄*ia-npb*,$st_i$] is a heuristic which only generates an
upper bound. Thus, its computation times are not comparable with the ones of DACGP.
However, we can compare the quality of the solutions found as well as the number of
problems solved. DACGP is coded in FORTRAN, B&B[1⋄*ia-npb*] and GA[1⋄*ia-npb*,$st_i$]
are coded in C. DACGP was run on an IBM-PS2 Model 80 PC (80386 processor) with
mathematical coprocessor 80387, we implemented B&B[1⋄*ia-npb*] on the same machine to
make computation times comparable.

Computational results for the DACGP are reported only for identical holding costs $h_i = 1$
for all items. Consequently, we solve DLSP as BSP(DLSP) and thus only need to consider
regenerative schedules, cf. Theorem 2.8.

The DLSP instances with nonzero setup times are provided by the authors of [24]. They
generated problems for item-period combinations $(N, T) = (2, 20), (2, 40), (4, 40), (2, 60)$,
$(4, 60), (6, 60)$. We refer only to problems with $T = 60$, because smaller problems are
solved much faster by B&B[1⋄*ia-npb*] than by DACGP. The DLSP instances have setup
times $st_{g,i}$ of either 0, 1 or 2 periods. The average setup-time per item (over all instances)
is (approximately) 0.5, making setup times not very significant. For each item-period
combination there are instances with different (approximate) capacity utilization $\rho$: low
(L) capacitated ($\rho < 0.55$), medium (M) ($0.55 \leq \rho < 0.75$) and high (H) capacitated
problems ($\rho \geq 0.75$). Approximate capacity utilization is defined as $\rho = 1/T \sum_{i,t} q_{i,t}$.
There are 30 instances for each $(N, T, \rho)$ combination, so that we have $3 \cdot 3 \cdot 30 = 270$
instances in total.

---

[5]In [24], the slack in a production schedule for an item $i$ corresponds to unsatisfied demand for this
item.

[6]This procedure is not further specified in [24]. In an earlier working paper a reference to the mathematical programming software LINDO [87] is given.

**Table 4.6:** *Comparison of DLSP and BSP Algorithms for SISTSC*

| (N,T) | #J | ρ | DLSP DACGP $\triangle_{avg}$ | $R_{avg}$ | $\#_{inf}$ | BSP GA[1◇ia-npb,$st_i$] $A_{avg}$ | $A_{max}$ | $\#_{inf}$ | B&B[1◇ia-npb] $R_{avg}$ | $\#_{inf}$ |
|---|---|---|---|---|---|---|---|---|---|---|
| (2,60) | 19 | L | 0.17 | 25.8 | 2 | 0.07 | 1.4 | 2 | 0.1 | 2 |
| | 25 | M | 0.20 | 76.3 | 7 | 0.23 | 2.6 | 7 | 0.2 | 7 |
| | 29 | H | 1.22 | 274.9 | 10 | 0.56 | 4.3 | 9 | 0.1 | 9 |
| (4,60) | 21 | L | 0.15 | 38.9 | 3 | 0.14 | 2.6 | 3 | 4.2 | 3 |
| | 31 | M | 0.47 | 120.8 | 6 | 0.63 | 5.8 | 5 | 11.8 | 5 |
| | 35 | H | 1.43 | 268.7 | 11 | 0.82 | 6.3 | 10 | 7.1 | 10 |
| (6,60) | 22 | L | 0.13 | 56.2 | 1 | 0.11 | 1.0 | 1 | 36.9 | 1 |
| | 33 | M | 0.70 | 264.9 | 10 | 0.38 | 4.5 | 7 | 149.0 | 7 |
| | 35 | H | 0.99 | 274.1 | 10 | 0.41 | 6.6 | 10 | 98.7 | 10 |

(386 PC with coprocessor)

In Table 4.6 we use #J to denote the average number of jobs in BSP(DLSP) for the problem size $(N, T)$ of the DLSP. For DACGP we use $\triangle_{avg}$ to denote the average gap (in percent) between upper and lower bound. $\#_{inf}$ is the number of problems found infeasible by the different procedures, and $R_{avg}$ denotes the average time (in seconds) needed for the 30 instances in each class. For DACGP, all values are taken from [24]. For GA[1◇ia-npb,$st_i$], we denote by $A_{avg}$ ($A_{max}$) the average (maximal) deviation from the optimal solution computed with B&B[1◇ia-npb].

Comparing between DACGP and B&B[1◇ia-npb], the B&B algorithm solves problems with $N = 2$ and $N = 4$ much faster; the number of sequences to examine is relatively small. For $N = 6$ computation times of B&B[1◇ia-npb] are comparable to DACGP. In (6,60,M) the simplex based procedure in DACGP finds a feasible integer solution for one of the 10 instances claimed infeasible in Table 4.6. Thus, in (6,60,M) 9 instances remain unsolved by DACGP, whereas with B&B[1◇ia-npb] only 7 instances are infeasible. DACGP also fails to find existing feasible solutions in problem classes $(N, T, \rho)$ =(2,60,H), (4,60,M). Recall that B&B[1◊*ia-npb*] takes advantage of a small solution space, keeping the enumeration tree small and thus detecting infeasibility or a feasible solution rather quickly. DACGP tries to improve the lower and upper bound, which is difficult without an initial feasible solution. Therefore the (heuristic solution procedure) DACGP may miss feasible solutions if the solution space is small.

From Table 4.6 we note that the deviations $A_{avg}$ and $\triangle_{avg}$ for DACGP and GA[1◊*ia-npb*,$st_i$] do not differ significantly, so the solution quality of both heuristics is approximately the same.[7] But note that GA[1◊*ia-npb*,$st_i$] finds all solutions in (6, 60,M) while DACGP fails to find a solution for some of the feasible instances. A comparison of the computation times between DACGP and GA[1◊*ia-npb*,$st_i$] is left out because DACGP also generates a lower bound. Moreover, computation times for a genetic algorithm directly depend on the choice of the parameters. For the results in Table 4.6 we choose a population size of 100 and maximal number of generations (=iterations) of 500. The genetic algorithm stops if 100 consecutive generations do not improve the solution.[8] Good results for GA[1◊*ia-npb*,$st_i$] can be obtained also for smaller populations and less iterations but then GA[1◊*ia-npb*,$st_i$] sometimes fails to solve instances where a feasible solution is difficult to find.

For the same problem size $(N, T)$ in DLSP the number of jobs $J$ in BSP(DLSP) may be very different. Therefore, solution times differ considerably for B&B[1◊*ia-npb*]. Table 4.7 presents the frequency distribution of solution times. In every problem class the majority of instances is solved in less than the average time $R_{avg}$ for DACGP.

## 4.3.2   Sequence Dependent Setup Costs (SDSC)

An algorithm to solve SDSC is proposed by Fleischmann [50]. Fleischmann transforms the DLSP into a traveling salesman problem with time windows (TSPTW), where a *tour* corresponds to a production schedule. He calculates a lower bound by lagrangean relaxation where the condition that each node is visited exactly once is relaxed. An upper bound is calculated by a heuristic. The heuristic procedure first constructs a tour for the

---

[7]The solutions of DACGP are not known, hence we cannot calculate $A_{avg}$ for DACGP.

[8]For these parameters, average (maximal) CPU seconds vary between 3.1 (4.8) and 11.9 (24.9) on a 486/33 Mhz PC.

**Table 4.7:** *Frequency Distribution of Solution Times of B&B[1◇ia-npb]*

| $(N,T)$ | | Number of instances solved faster than ... [sec] | | | | | | | | |
|---------|---|--------|------|-------|-------|--------|--------|---------|--------|-----------|
| | | < 0.1 | < 1 | < 10 | < 30 | < 100 | < 300 | < 1000 | ≥ 1000 | < $R_{avg}$ |
| $(2,60)$ | L | 10 | 20 | | | | | | | 30 |
| | M | 4 | 26 | | | | | | | 30 |
| | H | 13 | 17 | | | | | | | 30 |
| $(4,60)$ | L | | 4 | 23 | 3 | | | | | 30 |
| | M | | 2 | 16 | 10 | 2 | | | | 30 |
| | H | 3 | 4 | 13 | 10 | | | | | 30 |
| $(6,60)$ | L | 1 | | 4 | 5 | 10 | 8 | 2 | | 25 |
| | M | 1 | 1 | 3 | 5 | 12 | 5 | 2 | 1 | 26 |
| | H | 1 | | 4 | 5 | 10 | 8 | 2 | | 28 |

(386 PC with coprocessor)

TSPTW (and thus a solution of DLSP) and tries to improve the solution via an Or-opt operation. In Or-opt, pieces of the initial tour are exchanged to obtain an improved solution. Or-opt is repeated until no further improvements are found. We refer to Fleischmann's algorithm as TSPOROPT. TSPOROPT was coded in Fortran, experiments are performed on a 486DX2/66 PC with the code provided by Fleischmann.

Fleischmann divides the time axis into micro and macro periods. Holding costs occur only between macro periods, and demand occurs only at the end of macro periods. Thus a direct comparison of TSPOROPT and B&B[1◇ia-npb] is not made on Fleischmann's instances, but rather on randomly generated BSP instances which are transformed into DLSP instances. We generated 30 instances for $N = 5$ families and low (L) ($\rho \approx 0.75$) or high (H)($\rho \approx 0.97$) capacity utilization. Note that for zero setup times, $\rho$ does not depend on the schedule; the feasibility problem is polynomially solvable. In BSP we have an average number $\#J = 33$ of jobs with a processing time out of $DU(1,4)$. In DLSP we have an average $T = 73$ for high (H) and $T = 100$ for low (L) capacitated instances. Holding costs are identical, and we solve BSP(DLSP). From [50] we select the two setup

**Table 4.8:** *Comparison of DLSP and BSP Algorithms for SDSC*

| setup cost | | | | TSPOROPT | | | B&B[$1\diamond ia$-$npb$] |
|---|---|---|---|---|---|---|---|
| matrix | $(N,T)$ | #$J$ | $\rho$ | $\triangle_{avg}$ | $\triangle Z_{best}$ | $R_{avg}$ | $\tilde{R}_{avg}$ |
| S4 | $(5, 75)$ | 33 | H | 5.2 | 4.1 | 3.8 | 0.1 |
|    | $(5, 100)$ | 33 | L | 15.9 | 15.3 | 38.2 | 24.6 |
| S6 | $(5, 75)$ | 33 | H | 3.7 | 2.0 | 3.4 | 0.2 |
|    | $(5, 100)$ | 33 | L | 39.2 | 15.8 | 33.9 | 107.8 |

(486DX2/66 PC)

cost matrixes S4 and S6 which satisfy the triangle inequality: in .S4 costs equal 100 for
$g < i$ and 500 for $g > i$ while the setup structure is similar to $sq$. In S6 we have only two
types of setups: items $\{1, 2, 3\}$ and $\{4, 5\}$ form two setup-groups, with minor setup costs
of 100 within the setup-groups and major setup costs of 500 from one setup-group to the
other; this structure is similar to $gp$.

In Table 4.8 results are aggregated over the 30 instances in each class. We use $\triangle_{avg}$ to
denote the average gap between lower and upper bound (measured in %) for TSPOROPT
and $R_{avg}$ ($\tilde{R}_{avg}$) to denote the average time for TSPOROPT (B&B[$1\diamond ia$-$npb$]) in seconds.
We denote by $\triangle Z_{best}$ the average deviation in % of the heuristic solution of TSPOROPT
from the exact solution of B&B[$1\diamond ia$-$npb$]. Table 4.8 shows that $\triangle_{avg}$ can be quite large
for TSPOROPT. Solution times of B&B[$1\diamond ia$-$npb$] are short for high capacitated instances
and long for low ones. For S4 TSPOROPT generates a very good lower bound, we have
$\triangle_{avg} \approx \triangle Z_{best}$, and the deviation from the optimal objective is due to the poor heuristic
upper bound.[9] On the other hand, for S6 both the lower and the upper bound are not very
close to the optimum. It should be noted that B&B[$1\diamond ia$-$npb$] does not solve large instances
of SDSC with 8 or 10 items whereas Fleischmann reports computational experience for
instances of this size as well. The feasibility bound is much weaker for zero setup times,
or, equivalently, the solution space is much larger rendering B&B[$1\diamond ia$-$npb$] less effective.

---

[9]A similar observation has been made by Fleischmann who conjectures that *"primarily the heuristics
need to be improved"*([50], p. 403).

For the instances in Table 4.8, however, B&B[1⋄$ia$-$npb$] yields a better performance.


### 4.3.3  Sequence Dependent Setup Times and Costs (SDSTSC)

In Salomon et al. [110], Fleischmann's reformulation of the DLSP into a TSP with time windows (TSPTW) is extended to nonzero setup times in order to solve SDSTSC. Nodes in the TSP network represent positive demands, and all nodes must be visited in a certain time window. The reformulated DLSP is solved with a dynamic programming approach designed for TSPTW problems (cf. Dumas et al. [45]), we refer to the procedure in [110] as TSPTWA. Paths in the TSP network correspond to partial schedules. Similar to the dominance rule of Theorem 3.3, in TSPTWA paths may dominate other paths via a cost dominance, or they may be eliminated because they cannot be extended, which corresponds to the feasibility bound in Theorem 3.1.

TSPTWA is coded in C and run on a HP9000/730 workstation (76 mips, 22 M flops). B&B[1⋄$ia$-$npb$] runs on a 486DX2/66 PC.

To test TSPTWA, Salomon et al. [110] use randomly generated instances in which, similar to [24], setup times $st_{g,i} \in \{0; 1; 2\}$. Unfortunately, the setup times do not satisfy the triangle inequality. A "triangularization" (e.g. with the Floyd/Warshall algorithm) often results in setup times equal to zero. So we adjusted the setup times "upwards" (which possible in this case because $st_{g,i} \in \{0; 1; 2\}$) and as a result, setup times are mostly nonzero. We added 4 (8) units to the planning horizon for $N = 3$ and $N = 5$ ($N = 10$) in order to obtain the same (medium) capacity utilization as in [110]. In this way, the instances are supposed to have the same degree of difficulty for TSPTWA and B&B[1⋄$ia$-$npb$]: the smaller solution space from correcting $st_{g,i}$ upwards is compensated by a longer planning horizon.

In [110] instances are generated for $T = 20, 40, 60$, and we take the (largest) instances for the item-period combination $\{(N, T)\} = \{(10, 40); (3, 60); (5, 60); (10, 60)\}$. These instances have a medium (M) capacity utilization $0.5 \leq \rho \leq 0.75$ because setup times are nonzero. For each $(N, T)$ combination, 30 instances with and without holding costs are generated. Holding costs differ among the items. Consequently, we solve BSPUT(DLSP)

**Table 4.9:** *Comparison of DLSP and BSP Algorithms for SDSTSC*

| $(N,T)$ | $\#J$ | $h_i$ | TSPTWA | | B&B[$1\diamond ia$-$npb$] | | | |
|---|---|---|---|---|---|---|---|---|
| | | | $\#F$ | $R_{avg}$ | $\#\tilde{F}$ | $\tilde{R}_{avg}$ | $\tilde{R}^B_{avg}$ | $\triangle Z^B_{max}$ |
| (10, 48) | 25 | $= 0$ | 21 | $< 1200$ | 30 | 3.5 | – | – |
| | 25 | $> 0$ | 1 | $< 1200$ | 21 | (489.8) | (373.1) | 0.0 |
| (3, 64) | 38 | $= 0$ | 30 | $< 1200$ | 30 | 0.8 | – | – |
| | 38 | $> 0$ | 30 | $< 1200$ | 30 | 5.1 | 1.4 | 0.4 |
| (5, 64) | 38 | $= 0$ | 25 | $< 1200$ | 30 | 28.0 | – | – |
| | 38 | $> 0$ | 0 | | 30 | 140.3 | 36.3 | 0.0 |
| (10, 68) | 38 | $= 0$ | 4 | $< 1200$ | 14 | (41.6) | – | – |
| | 38 | $> 0$ | 0 | | 2 | (438.6) | (488.4) | 0.0 |

(486DX2/66 PC)

if $h_i > 0$ and BSP(DLSP) for $h_i = 0$. Again, in the latter case, we only need to consider regenerative schedules. In Table 4.9, $\#F$ ($\#\tilde{F}$) denotes the number of problems solved by TSPTWA (B&B[$1\diamond ia$-$npb$]) within a time limit of 1200 sec (1200 sec) and a memory limit of 20 MB (10 MB). $\#J$ denotes the average number of jobs for the BSP. $\tilde{R}_{avg}$ ($\tilde{R}^B_{avg}$) denotes the average time required by B&B[$1\diamond ia$-$npb$] to solve the instances (considering only regenerative schedules). The average time is calculated over all instances which are solved within the time limit, $\tilde{R}_{avg}$ is put in brackets if not all instances are solved. The last column shows the results obtained from considering only regenerative schedules during the enumeration for $h_i > 0$: $\triangle Z^B_{max}$ provides the maximal deviation in % from the optimal solution (which may be non-regenerative).

Table 4.9 demonstrates that B&B[$1\diamond ia$-$npb$] succeeds in solving some of the problems which remained unsolved by TSPTWA. Solution times of B&B[$1\diamond ia$-$npb$] are short compared with TSPTWA for $N = 3$ and $N = 5$. Solution times increase for $N = 10$, and instances can only be solved if the number of jobs is relatively small. Instances are difficult for nonzero, and so especially for non-identical holding costs. If we only enumerate over regenerative schedules, solution times for B&B[$1\diamond ia$-$npb$] decrease. Moreover, only one instance is not solved to optimality for $(N, T) = (3, 64)$. Thus, even for non-identical holding costs optimal solutions are regenerative in most cases. Furthermore, for $(N, T) = (10, 48)$ $((10, 68))$, 29 (3) instances would have been solved within the time limit of 1200 sec considering only regenerative schedules.

### 4.3.4   Single-Level Batch Availability (BA)

Brüggemann and Jahnke [19] propose a simulated annealing procedure for BA; we refer to their procedure as BRJSA. The neighborhood in their simulated annealing approach is based on the representation of a DLSP in the $Y_{i,t}$ matrix: starting with an initial solution, where cumulated production equals cumulated demand, two columns in the matrix are interchanged to obtain a neighborhood solution. This way capacity restrictions per period are not violated, and total production for each item remains the same. However, if the new solution does not satisfy all demands $q_{i,t}$ in time, for each item a certain initial and end inventory level is needed which then incurs corresponding penalty costs. The simulated annealing algorithm BRJSA seeks to minimize the sum of holding, setup and penalty costs.

Setup times and setup costs are sequence independent in [19], and BRJSA is compared with B&B[1◇ba] and GA[1◇ba,st$_i$]. In [19], computational experience is reported for one hard instance (hard to find a feasible solution) with $N = 6$ items and $T = 60$ periods, which takes 1204 sec to solve on a 486/20 PC. We solve the same instance on a 486DX2/66 PC with GA[1◇ba,st$_i$] (B&B[1◇ba]) in 10 (10) sec. A non-late schedule (cf. Definition 2.10) is optimal for this instance because the genetic algorithm finds a better solution than the B&B algorithm. The sequence of batches and also the objective function value are slightly different in our solution than the one displayed in [19]. However, a *feasible* solution is found in much less time.

### 4.3.5   Two-Level Batch Availability (BA2LV)

In [20] Brüggemann and Jahnke extend their simulated annealing approach in order to solve BA2LV; we refer to their procedure as BRJSA2LV. Similar to the genetic algorithm of Section 3.4.4, BRJSA2LV solves the problem on the second level first, and the solution

**Table 4.10:** *Comparison of DLSP and BSP Algorithms for BA2LV*

| | | | BRJSA2LV | | GA$[F\diamond ba,st_i]$ | | |
|---|---|---|---|---|---|---|---|
| $(N,T)$ | $\#J$ | $\rho$ | $R_{avg}$ | $\#UT$ | $5 \cdot R_{avg}$ | $\#UT$ | $|PS|$ |
| $(6,60)$ | 22 | 60-70 | 356 | - | 20 | - | 50 |
| | 22 | 70-80 | 351 | - | 20 | - | 50 |
| | 22 | 80-85 | 400 | - | 20 | - | 50 |
| | 22 | 85-88 | 492 | 2 | 20 | - | 50 |
| | 22 | 88-92easy | 670 | - | 40 | 2 | 100 |
| | 22 | 88-92hard | 1119 | 29 | 40 | 6 | 100 |

(486DX2/66 PC)

of the second level becomes the demand matrix for the first level. The two-level DLSP corresponds to problem $[F/fam,ba,st_i,d_{(i,j)}/\sum w_{(i,j)}C_{(i,j)}+\sum sc_i]$, and we can only compare the two heuristic approaches GA$[F\diamond ba,st_i]$ and BRJSA2LV (cf. Table 4.10).

Brüggemann and Jahnke give computational experience for instances with $N = 6$ items and $T = 60$ periods in 6 problem classes with 15 randomly generated instances each. Approximate capacity utilization $\rho$ is given in % for each class, e.g. $\rho$ is between 60 and 70 % in the first class. For high capacity utilization, i.e. $\rho > 0.88$, they differentiate between a hard and an easy problem class. We choose a similar experimental design as Brüggemann and Jahnke and repeat GA$[F\diamond ba,st_i]$ 10 times (=10 trials) for each instance with a rather small population size $|PS|$. So GA$[F\diamond ba,st_i]$ may come up with infeasible solutions in some of the 10 trials, which also happened with BRJSA2LV. For both algorithms, we denote the total number of unsuccessful trials in a problem class by $\#UT$.[10] For BRJSA2LV experiments are conducted on a 486/20 PC, for GA$[F\diamond ba,st_i]$ on a faster 486DX2/66 PC, so we multiplied times on the latter with factor 5. Table 4.10 gives average running times $R_{avg}$ ($5 \cdot R_{avg}$) for one trial in seconds for BRJSA2LV (GA$[F\diamond ba,st_i]$). The different population sizes used for the genetic algorithm are denoted by $|PS|$.

---

[10]Cf. e.g. Table 4.10 for problem class 88-92hard. The number of unsuccessful trials $\#UT$ is 29 (6) out of $10 \cdot 15$ trials for BRJSA2LV (GA$[F\diamond ba,st_i]$).

Table 4.10 shows that the running times for GA[$F\diamond ba,st_i$] are considerably smaller than those for BRJSA2LV in all problem classes. Like the results in [20], the objective function values of GA[$F\diamond ba,st_i$] differ only slightly over the 10 trials, so we conclude that the algorithm has converged to a "good" solution. But more important than the comparison of objective function values is the ability of GA[$F\diamond ba,st_i$] to find a feasible solution faster than BRJSA2LV.[11]

## 4.4 Conclusions

The study in this chapter compares a lotsizing with a scheduling model (and the algorithms for each one of the models) for the same planning problem – in contrast to a study which compares (only) the performance of different algorithms for the same model. Therefore, the following remarks address not only the algorithmic performance but the "suitability" of a model as well. Our comparison is restricted to the DLSP, as there is something like a "standard" DLSP model well established in literature.

Another important lotsizing and scheduling model is the proportional lotsizing and scheduling problem (PLSP), proposed by Haase [61] and extended by Kimms [75] to the multi-level case on one machine. In the PLSP, the setup state is kept up over idle periods, which corresponds to the batching type *ia-pb*. We conjecture that a similar comparison of the PLSP with the BSP with *ia-pb* is also possible.[12]

Concerning the comparison between the **models**, the basic difference is a decision based on periods (discrete time intervals) in the DLSP versus the construction of a schedule

---

[11]Objective function values can only be compared for an example given in [20], (Table 1 and Figures 3 to 5). In [20] a solution is given for the second level, which consists of a sequence of 10 batches. This solution can be improved interchanging the positions of the batch 2 with 3 and batch 6 with 7, which is the solution of GA[$F\diamond ba,st_i$] on the second level for this instance. In [20] batches 2,3, and 6,7 are not ordered according to Theorem 2.10.

[12]In the PLSP there is no unit capacity per period, and not more than two items can be produced per period. The construction of an equivalent BSP with these assumptions requires much more technical overhead and modification. However, for the models in Chapter 2 an equivalence can be stated for a special case: the PLSP with demands which are multiples of the (time invariant) period capacity is equivalent to [$1/fam, ia\text{-}pb,st_{g,i},d_{(i,j)}/ \sum w_{(i,j)}C_{(i,j)} + \sum sc_{g,i}$] if holding costs are equal.

in the BSP (continuous time); the fact which makes both models comparable is that they generate corresponding solutions.  The decisions in each period are subject to the capacity constraint per period in lotsizing, and the construction of a schedule is subject to the disjunction between jobs.  But both are basically the same constraint: the machine can produce only one family at a time.[13]  We furthermore observe that the problem size $N \times T$ for DLSP can be quite different from the problem size $J$ (number of jobs) for BSPUT(DLSP) and BSP(DLSP).  For another segmentation of the planning horizon $T$, the problem size of DLSP increases, but remains the same for BSP(DLSP).  In the BSP, all parameters can also be real numbers: setup times, in particular, are not restricted to being multiples of a period length as in the DLSP.

For the different DLSP variants many parameters are associated with the sequence, in particular for sequence dependent setups.[14]  Then, a *job sequence* is the main characteristic of a solution, and the BSP takes that into account.  The BSP decision variables "better" characterize the solution than the DLSP decision variables.  Consequently, solving the planning problem as a BSP is easier, as indicated by the comparison in Section 4.3.

On the other hand, if many parameters are associated with the period $t$, e.g. time varying production costs or capacity, and if production lots are not linked over consecutive periods, the period is the planning entity.  Then, a solution is mainly characterized by *the decision for each individual period* and a lotsizing model is better suited.

Regarding **algorithmic** comparison, we observe that one algorithm for the BSP, B&B[1◇ia-npb], compares favorably with the (specialized) algorithms DACGP, TSPORPT and TSP-TWA for the DLSP variants.  For some variants, exact solutions are found in less time than heuristic solutions.  However, the algorithm for the DLSP with sequence *independent* setup costs and zero setup times in Fleischmann [49] solves larger problem sizes than those

---

[13]In multi-level problems the constraints imposed by the product structure are also expressed in different ways in scheduling and in lotsizing models. On the one hand there are *precedence constraints* between jobs for the construction of a schedule, and planning is done for *jobs*. On the other hand there are *inventory balance constraints* in lotsizing, and planning is done for *quantities*.

[14]As a consequence, the mathematical programming formulation in Table 4.3 becomes more difficult in this case. A much shorter formulation can be given for the (standard) DLSP with zero setup times and sequence independent setup costs (cf. e.g. Fleischmann [49]).

which can still be solved by B&B[1◇*ia-npb*].

When solving the BSP with a B&B algorithm to optimality, we face the difficulty that the feasibility problem itself is already difficult. We must maintain feasibility and minimize costs at the same time. Compared with other scheduling models, the objective function is quite difficult for the BSP. An easily calculable, tight lower bound could thus not be developed. We therefore use dominance rules to prune the search tree. Not surprisingly, the smaller the solution space (i.e. long setup times, high capacity utilization), the better the performance of the enumerative algorithm. In the case of batch availability, the effects of batching are emphasized. The genetic algorithm, which tackles batching separately, compares favorably with algorithms for the DLSP – in particular, if a *feasible* solution is difficult to find.

To summarize, the BSP algorithms are superior if the solution space is small, but especially in this case decision support for short term planning is needed. On the other hand, a large solution space suggests mathematical programming based approaches, such as DACGP or the algorithm proposed by Fleischmann [49].

# Chapter 5

# Methods for the Multi-Level and Parallel Machine Case

In this chapter we extend the single-level single-machine case to the multi-level case as well as to the case of parallel identical machines. Using the three-field notation in Table 2.2, we consider problems $[\alpha/fam,ia\text{-}pb,st_{g,i},d_{(i,j)}/\sum w_{(i,j)}C_{(i,j)} + \sum sc_{g,i}]$ where $\alpha \in \{ML1, ML, P\}$. For these problems, results from the single-machine case are extended. We restrict the analysis to the batching type $ia\text{-}pb$ as then, essentially, we only need to modify the branch and bound algorithm B&B$[1 \diamond ia\text{-}pb]$ described in Section 3.2. There are many heuristic methods for the multi-machine case (cf. e.g. Brucker et al. [17], Gupta [58], Monma and Potts [93], Sotskov et al. [115], Tang [118]), but there is a lack of exact algorithms (with the exception of Schutten and Leussink [113]). Consequently, our focus is on an exact algorithm, which is useful to solve benchmark instances for heuristics, and the major contribution is the extension of the single-machine enumeration scheme to the multi-level and parallel machine case.

In the sequel we introduce the extended notation and illustrate a rightshift-rule based on Theorem 2.3. Section 5.1 covers the multi-level case, where we refer to the model in Section 2.4. The multi-level single-machine case is examined in Section 5.1.1, the multi-machine case with a fixed family-machine assignment in Section 5.1.2. Preliminary computational results follow in Section 5.1.3. We treat the case of parallel identical ma-

**Table 5.1:** *Attributes of Partial Schedules III*

| | |
|---|---|
| $(i^s, j^s)$ | job under consideration at stage $s$ |
| $m^s$ | the machine $m$ on which $(i^s, j^s)$ is scheduled |
| $i^s_m$ | the $s$-partial schedule $\sigma^s$ starts with family $i^s_m$ on machine $m$ |
| $t_m(\sigma^s)$ | start time of the $s$-partial schedule $\sigma^s$ on machine $m$ |
| $c(\sigma^s)$ | costs of $\sigma^s$ |
| $\mathcal{AS}^s_m (\mathcal{US}^s_m)$ | set of jobs already scheduled (unscheduled) on machine $m$ in the $s$-partial schedule $\sigma^s$ |
| $\mathcal{UI}^s_m$ | set of families to which jobs in $\mathcal{US}^s_m$ belong |

chines in Section 5.2. Again, some preliminary computational results follow at the end of the section. For the algorithms, we report computational experience with instances constructed by the generator in Section 2.7; procedures in literature are not available for a comparison.

The B&B algorithm uses the basic enumeration scheme of Section 3.1. Semiactive schedules are optimal for *ia-pb* (cf. Theorem 2.2). Consequently, we (implicitly) enumerate all *sequences* for the multi-level and all *sequences* and *machine assignments* for the parallel machine case. In contrast to this approach, Schutten and Leussink [113]) schedule each job on the machine on which it is completed first (for the maximum lateness objective) and only need to enumerate the sequences. For the BSP a similar approach does not work for $\gamma = *$, but, for instance, for $\gamma = \sum w_{(i,j)} C_{(i,j)}$, cf. Kurth [78].

Table 5.1 summarizes the attributes of partial schedules; note that some of the attributes must now be indexed with $m$. At stage $s$, we consider job $(i^s, j^s)$ and schedule it on machine $m^s$. Costs of a partial schedule $\sigma^s$ are denoted by $c(\sigma^s)$. $\sigma^s$ denotes the whole schedule on all machines, the start times and the last scheduled families of $\sigma^s$ on the different machines $m$ are denoted by $t_m(\sigma^s)$ and $i^s_m$, respectively. For $\alpha = ML$, we can determine the set of jobs to be scheduled on each machine $m$, and can thus identify the set of unscheduled jobs $\mathcal{US}^s_m$ for each machine $m$ separately.

The number of sequences to be examined in the multi-machine case is significantly reduced

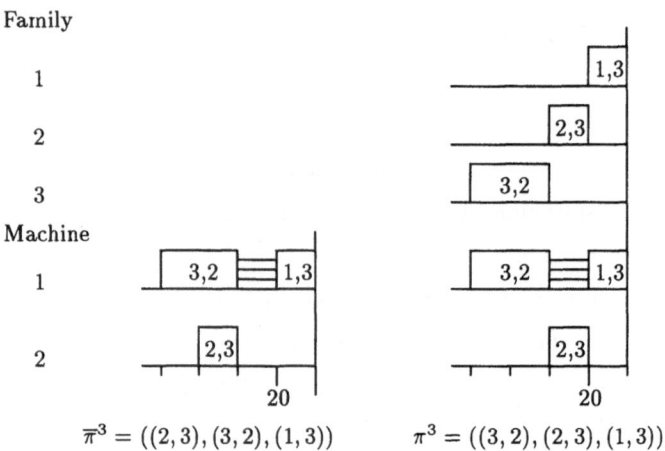

**Figure 5.1:** *Rightshift Rule for Multi-Machines*

by a **rightshift-rule** due to Theorem 2.3. This rule allows to fathom a partial sequence if in the corresponding schedule job $(i^s, j^s)$ can be rightshifted.

Consider Figure 5.1 for an illustration based on the example in Figure 2.5 (p. 30): we only consider the jobs (1,3), (2,3) and (3,2) with processing times 1, 1 and 2 and deadlines 21, 20 and 19, respectively, which are depicted in the upper part of Figure 5.1. From Theorem 2.3 we know that there is an optimal schedule where completion times increase monotonically such that $C_{(i_{[k]}, j_{[k]})} \leq C_{(i_{[k+1]}, j_{[k+1]})}$. Consider sequence $\overline{\pi}^3 = ((2,3), (3,2), (1,3))$. Scheduling each job rightmost with monotonically increasing completion times, we obtain $\overline{\sigma}^3 = (19, 19, 21)$ because $C_{(2,3)} = 19 \leq C_{(3,2)} = d_{(3,2)} = 19$. $\overline{\sigma}^3$ is not semiactive (and thus not optimal), the semiactive schedule is $\tilde{\sigma}^3 = (20, 19, 21)$ (where job $(\overline{i^3}, \overline{j^3})=(2,3)$ is *rightshifted*), but now completion times do not increase monotonically. Thus, sequence $\overline{\pi}^3$ can be fathomed by Theorem 2.3.

Consider $\pi^3 = ((3,2), (2,3), (1,3))$, there is a semiactive schedule $\sigma^3 = (19, 20, 21)$ and now completion times increase monotonically.[1]

---

[1] If we consider only semiactive schedules where completion times do *not* increase monotonically, *different* sequences would lead to the *same* schedule: in our example $\tilde{\sigma}^3 = (20, 19, 21)$ for $\overline{\pi}^3 =$

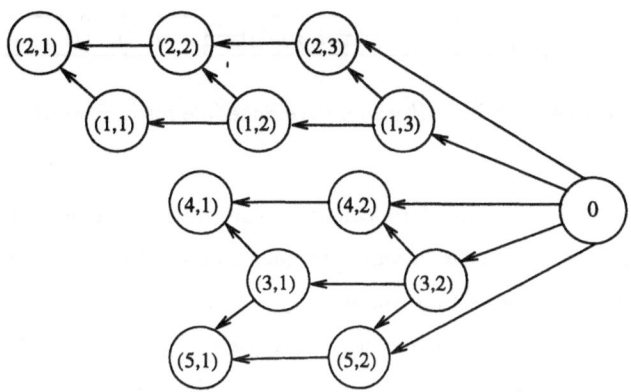

**Figure 5.2:** *Precedence Graph in the Multi-Level Case*

## 5.1 Multi-Level Case

In the multi-level case, the EDDWF precedence graph contains additional arcs. Figure 2.6 (p. 31) displays the precedence constraints between jobs which are due to the gozinto graph given in the example of Section 2.4. Jobs are in EDDWF order, and we employ a backward scheduling scheme (so that arcs are reverted w.r.t. their orientation in Figure 2.6); the corresponding precedence graph is depicted in Figure 5.2. Compared with the single-level case in Figure 3.1 (p. 55), the precedence graph in Figure 5.2 establishes a "stronger" partial ordering; due to the gozinto graph there are additional arcs between jobs of different families.

### 5.1.1 Single-Machine

The algorithm for the single-machine case ($\alpha = ML1$) is denoted by B&B[$ML1\diamond ia\text{-}pb$], and it differs from B&B[$1\diamond ia\text{-}pb$] only in its precedence graph. So we have the same feasibility and cost bounds and thus can apply the same dominance rule, cf. Theorem 3.3.

$((2,3),(3,2),(1,3))$ and $\sigma^3 = (19,20,21)$ for $\pi^3 = ((3,2),(2,3),(1,3))$ denote the same schedule for different sequences. So the rightshift-rule curtails the enumeration over sequences, hence each schedule is generated only once.

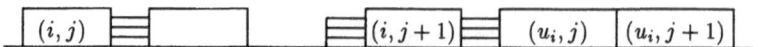

**Figure 5.3:** *Non-Regenerative Schedule for* $\alpha = ML1$

Due to Theorem 2.8 we only need to consider regenerative schedules for intermediate families $i \in \mathcal{I}$ if holding costs are identical. Consider Figure 5.3 which illustrates a schedule which is non-regenerative: jobs $(u_i, j)$ and $(u_i, j + 1)$ are batched while the predecessor jobs $(i, j)$ and $(i, j + 1)$ form two batches. In Figure 5.3, job $(i, j + 1)$ starts a batch though there is still inventory of the (intermediate) family $i$. These schedules need not be considered in B&B[$ML1\diamond ia\text{-}pb$].

## 5.1.2   Multi-Machine

The algorithm for the multi-machine case is denoted by B&B[$ML\diamond ia\text{-}pb$]. Unfortunately, for $\alpha = ML$ we cannot restrict ourselves to regenerative schedules. The rightshift-rule in B&B[$ML\diamond ia\text{-}pb$] leads to a tighter feasibility bound as follows: for each machine $m$ we can calculate the minimum time $\mathsf{T}_m^s$ needed for a completion of $\sigma^s$ as

$$\mathsf{T}_m^s = \sum_{i \in \mathcal{UI}_m^s} \min_{\substack{g=0,\ldots,N \\ g \neq i}} \{st_{g,i}\} + \sum_{(i,j) \in \mathcal{US}_m^s} p_{(i,j)}.$$

We can now state that $\sigma^s$ has no feasible completion $\omega^s$ if $\exists m : \mathsf{T}_m^s > \min\{t_m(\sigma^s), C_{(i^s,j^s)}\}$. The second entry in the "min"-term is due to the rightshift-rule: if $\sigma^s$ has a feasible completion there must be an $\omega^s$ where all completion times are less or equal $C_{(i^s,j^s)}$. Feasibility is thus separately checked for each machine. The cost bound, however, is the same as in Theorem 3.2.

The dominance rule in Theorem 5.1 basically applies the single-machine dominance rule of Theorem 3.3 to each machine. Since jobs are assigned to machines, we know that if two partial schedules contain the same set of jobs, they also schedule the same set of jobs to each machine: for two partial schedules $\sigma^s$ and $\overline{\sigma}^s$ with $\mathcal{AS}^s = \overline{\mathcal{AS}}^s$ we also have $\mathcal{AS}_m^s = \overline{\mathcal{AS}}_m^s \; \forall m$.

**Theorem 5.1** *Consider two s-partial schedules $\sigma^s$ and $\overline{\sigma}^s$ with $\mathcal{AS}^s = \overline{\mathcal{AS}}^s$. $\sigma^s$ dominates $\overline{\sigma}^s$ if*

$$(i) \quad t_m(\overline{\sigma}^s) + st_{\overline{i}^s_m, i^s_m} \leq t_m(\sigma^s) \qquad \forall m,$$

$$(ii) \quad c(\overline{\sigma}^s) - \sum_{m=1}^{M} sc_{\overline{i}^s_m, i^s_m} \geq c(\sigma^s) \qquad and$$

$$(iii) \quad C_{(\overline{i}^s, \overline{j}^s)} \leq C_{(i^s, j^s)}.$$

**Proof:** The proof is the same as for Theorem 3.3, we apply Theorem 3.3 on each machine. Condition $(iii)$ is needed as we only consider schedules where completion times increase monotonically, cf. Theorem 2.3.

Let $\omega^s$ be a completion of $\overline{\sigma}^s$ so that $(\omega^s, \overline{\sigma}^s)$ is a solution. Let $\omega^s_m$ be the part of $\omega^s$ which is scheduled on machine $m$. $(\omega^s, \sigma^s)$ is also a solution as all parts $\omega^s_m$ complete $\overline{\sigma}^s$ on each machine $m$ due to $(i)$ and $(iii)$. Furthermore, $\omega^s$ completes $\sigma^s$ at lower cost because of $(ii)$, so that $\sigma^s$ dominates $\overline{\sigma}^s$. $\qquad \square$

## 5.1.3  Preliminary Computational Results

The complex multi-level model allows the analysis of numerous different factors in testing B&B[$ML1\diamond ia\text{-}pb$] and B&B[$ML\diamond ia\text{-}pb$], therefore only preliminary computational results are presented. Again we try to generate instances which are supposed to have no special structure, and therefore we restrict ourselves to a medium capacity utilization $\rho = M$ and a random setup structure $rd$. As for $\alpha = 1$, CPU times increase (decrease) for $\rho = L$ (H) so that we do not vary $\rho$ again.

In the **single-machine** case ($\alpha = ML1$), we examine the influence of different gozinto graphs on running times. The problem size is $(N, J) = (6, 36)$ for all instances; the number of families $N = 6$ is supposed to be large enough such that the different gozinto graphs exhibit different characteristics. We choose a linear structure (LIN), a general structure (GEN) and a convergent structure (CON) as depicted in Figure 5.4. In Table 5.2 we observe that computation times increase from LIN to CON. The partial ordering of jobs caused by LIN (GEN) is "stronger" than the one caused by GEN (CON), and consequently,

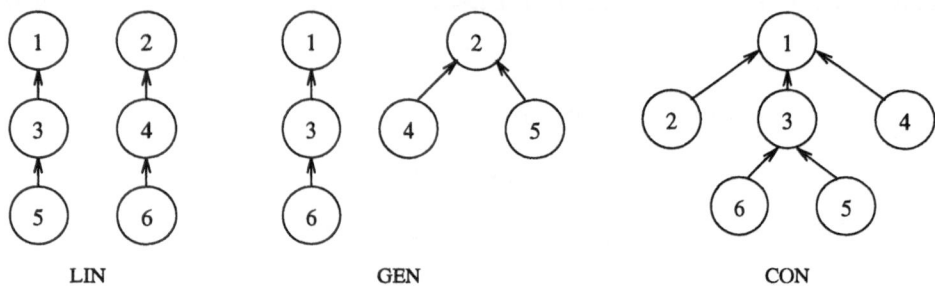

LIN                    GEN                    CON

**Figure 5.4:** *Gozinto Graphs of the Test Instances*

**Table 5.2:** *Preliminary Computational Results for* $\alpha = ML1$

| $(N, J) = (6, 36)$ | B&B[ML1$\diamond$*ia-pb*] | | | | | |
|---|---|---|---|---|---|---|
| | LIN | | GEN | | CON | |
| $\rho = M$ | $R_{avg}$ | $R_{max}$ | $R_{avg}$ | $R_{max}$ | $R_{avg}$ | $R_{max}$ |
| | 10.0 | 24.3 | 16.1 | 37.3 | 38.4 | 90.3 |

(IBM PowerPC)

the size of the enumeration tree grows from LIN to CON which explains the longer running times.

B&B[1$\diamond$*ia-pb*] without the occupation bound in Table 3.3 yields CPU times $(R_{avg}, R_{max}) = (51.6, 235.2)$ (for $\rho = $M and *st-struc* $= rd$). Compared with this running times[2] we can solve the same problem size $(N, J) = (6, 36)$ in a shorter time in the multi-level case because the gozinto graph curtails the enumeration tree.

In the **multi-machine** case $(\alpha = ML)$, only a considerable smaller problem size $(N, J) = (6, 24)$ could be solved. We vary the number of machines from $M = 2$ to $M = 3$, we consider two different gozinto graphs and two different machine assignments, resulting in $2 \cdot 2 \cdot 2 = 8$ problem classes. We take the gozinto graphs LIN and CON; LIN is supposed to cause a strong partial ordering of the jobs in the precedence tree, resulting

---

[2]And B&B[ML1$\diamond$*ia-pb*] does not use the occupation bound, either.

**Table 5.3:** *Machine Assignments $ma_i$ for $\alpha = ML$*

|  |  | LIN | | | | | | CON | | | | | |
|---|---|---|---|---|---|---|---|---|---|---|---|---|---|
|  | $i$ | 1 | 2 | 3 | 4 | 5 | 6 | 1 | 2 | 3 | 4 | 5 | 6 |
| $M = 2$ | SEP | 1 | 1 | 1 | 2 | 2 | 2 | 1 | 1 | 1 | 2 | 2 | 2 |
|  | INT | 1 | 2 | 2 | 1 | 1 | 2 | 1 | 2 | 2 | 1 | 1 | 2 |
| $M = 3$ | SEP | 1 | 1 | 2 | 2 | 3 | 3 | 1 | 1 | 2 | 2 | 3 | 3 |
|  | INT | 1 | 3 | 2 | 2 | 3 | 1 | 1 | 3 | 2 | 2 | 1 | 3 |

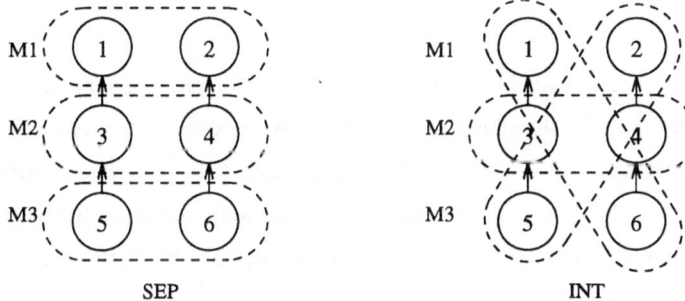

**Figure 5.5:** *Separated and Interdependent Machine Assignment for LIN*

in a smaller enumeration tree, whereas the partial ordering caused by CON is weaker and the enumeration tree is larger. Furthermore, we vary the machine assignment $ma_i$. A *separated* (SEP) machine assignment $ma_i$ suggests a decomposed approach machine by machine, so that $ma_i$ is flow-shop like. In an *interdependent* (INT) machine assignment, the first and last level may be assigned to the same machine. In INT, the machine assignments of the different levels are supposed to be strongly interdependent. In Figure 5.5 we provide an example for a separated and an interdependent machine assignment for LIN. For each of the 8 problem classes the family-machine assignments $ma_i$ are given in Table 5.3.

Table 5.4 displays average $R_{avg}$ and maximum $R_{max}$ running times in the 8 different problem classes. Enumeration is efficiently curtailed by the rightshift-rule, but the dominance

**Table 5.4:** *Preliminary Computational Results for* $\alpha = ML$

| $(N, J) = (6, 30)$ | | B&B[$ML\diamond ia\text{-}pb$] | | | |
|---|---|---|---|---|---|
| | | LIN | | CON | |
| $\rho = M$ | | $R_{avg}$ | $R_{max}$ | $R_{avg}$ | $R_{max}$ |
| $M = 2$ | SEP | 2.2 | 10.7 | 20.8 | 116.1 |
| | INT | 2.8 | 11.4 | 33.1 | 183.3 |
| $M = 3$ | SEP | 2.2 | 5.5 | 19.0 | 53.7 |
| | INT | 4.4 | 16.7 | 39.2 | 114.4 |

(IBM PowerPC)

rule in Theorem 5.1 is weaker (condition ($i$) must be valid $\forall m$), so that the solvable problem size is smaller than for the single-machine case. As in Table 5.2, running times increase from LIN to CON because the size of the enumeration tree increases. Furthermore, we observe that the machine assignment INT is more difficult to solve. Surprisingly at first sight, we find that running times do not *increase* significantly with a larger number of machines, i.e. from $M = 2$ to $M = 3$.[3] The rationale behind this finding is that for a fixed $N$, the larger the number of machines, the smaller the number of families assigned to the same machine, and hence, the number of permutations or setups between families $i$ does not increase from $M = 2$ to $M = 3$ in B&B[$ML\diamond ia\text{-}pb$].

## 5.2   Parallel Machines

The B&B algorithm for identical parallel machines is written as B&B[$P\diamond ia\text{-}pb$]; in this case, enumeration is performed over all EDDWF sequences *and* machine assignments. We still apply depth-first search, but in the case of backtracking we first branch to all possible machine assignments $MA_k$ of a job before considering another eligible job. In this way, the rightshift-rule fathoms most of the nodes in the backtracking steps. If we do not backtrack

---

[3]Recall, however, that the case $M = N$ is polynomially solvable: for each job $(i, j)$ of an intermediate family $i \in \mathcal{I}$ compute an artificial deadline $\tilde{d}_{(i,j)} = d_{(u_i,j)} - p_{(u_i,j)}$ and schedule all jobs in an EDD order.

but extend a partial schedule, we first branch to the job-machine combination $((i^s, j^s), m^s)$ where $(i^s, j^s)$ can be scheduled rightmost, i.e. we look for the pair $((i^s, j^s), m^s)$ where $C_{(i^s,j^s)}$ is maximum. This branching rule leads to good starting solutions and enhances the performance of B&B$[P\diamond ia\text{-}pb]$ significantly.

For the feasibility check, we compare the total time available $\mathsf{T}_{avlb}$ on all machines

$$\mathsf{T}_{avlb} = \sum_{m=1}^{M} \min\{t_m(\sigma^s), C_{(i^s,j^s)}\}$$

with the total time $\mathsf{T}^s$ needed to schedule the jobs in $\mathcal{US}^s$ (cf. Theorem 3.1).[4] Again, the second entry in the "min"-term takes the rightshift-rule into account. A partial schedule $\sigma^s$ cannot be extended to a solution if $\mathsf{T}^s > \mathsf{T}_{avlb}$. The cost bound is the same as in Theorem 3.2.

The dominance rule in the parallel machine case again compares two partial schedules $\sigma^s$ and $\overline{\sigma}^s$ which schedule the same set of jobs. But now, the set of jobs scheduled on different machines may be different and it is not clear which machines to compare. Therefore, we determine (heuristically) a *matching vector* $X = (X_1, \ldots, X_m, \ldots, X_M)$ defining a permutation of machines. By $X$ we compare the part of schedule $\overline{\sigma}^s$ on machine $X_m$ with the part of schedule $\sigma^s$ on machine $m$.

$\overline{\sigma}^s$ is dominated if for each $m = 1, \ldots, M$ the schedule on machine $m$ in $\sigma^s$ is "better" than the schedule on machine $X_m$ in $\overline{\sigma}^s$.

**Theorem 5.2** *Consider two s-partial schedules $\sigma^s$ and $\overline{\sigma}^s$ with $\mathcal{AS}^s = \overline{\mathcal{AS}}^s$. $\sigma^s$ dominates $\overline{\sigma}^s$ if $\exists\, X$ defining a machine permutation with*

$$(i) \quad t_{X_m}(\overline{\sigma}^s) + st_{\overline{i^s_{X_m}}, i^s_m} \leq t_m(\sigma^s) \qquad \forall m,$$

$$(ii) \quad c(\overline{\sigma}^s) - \sum_{m=1}^{M} sc_{\overline{i^s_{X_m}}, i^s_m} \geq c(\sigma^s) \qquad and$$

$$(iii) \quad C_{(\overline{i^s}, \overline{j^s})} \leq C_{(i^s, j^s)}.$$

---

[4]For $\alpha = P$ we may compute a tighter $\mathsf{T}^s$ as $\mathsf{T}^s = M \min_{i=1,\ldots,N} st_{0,i} + \sum_{(i,j)\in\mathcal{US}^s} p_{(i,j)}$ which is a tighter feasibility bound if the "start setups" $st_{0,i}$ are large.

Machine

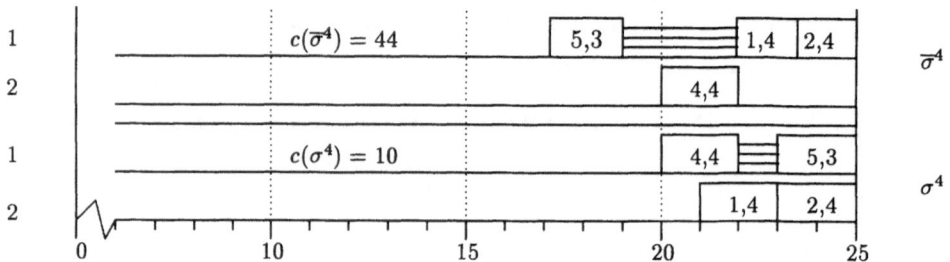

**Figure 5.6:** *Theorem 5.2 - $\sigma^4$ dominates $\overline{\sigma}^4$*

**Proof:** Let $\omega^s$ be a completion of $\overline{\sigma}^s$ so that $(\omega^s, \overline{\sigma}^s)$ is a solution. Let $\omega^s_{X_m}$ $(\sigma^s_m)$ be the part of $\omega^s$ $(\sigma^s)$ which is scheduled on machine $X_m$ $(m)$.

In the parallel machine case we do not necessarily have $AS^s_m = \overline{AS^s_{X_m}}$ for the pair of matched machines $(m, X_m)$. But each part $\omega^s_{X_m}$ can be scheduled on $m$ as a completion of $\sigma^s_m$ due to $(i)$ and $(iii)$ so that $(\omega^s, \sigma^s)$ is a solution, too. We need condition $(iii)$ due to the right-shift-rule. Thus, any completion $\omega^s$ of $\overline{\sigma}^s$ also completes $\sigma^s$, permuting the machines as in $X_m$.

Because of $(ii)$, $\omega^s$ completes $\sigma^s$ at lower costs, completing the proof.          $\square$

The heuristic for determining the vector $X$ first determines all "matchings", i.e. the pairs of machines where $\overline{i^s_{X_m}} = i^s_m$, the remaining components of $X$ are the "non-matched" machines in ascending order. Theorem 5.2 is stated for a given vector $X$; the *existence* of one $X$ is sufficient for the dominance rule in Theorem 5.2. However, the dominance rule is applied at each node and checked only for one $X$ determined (heuristically) as above. The primary purpose of the dominance rule is to rule out "symmetric" solutions, where $\sigma^s$ and $\overline{\sigma}^s$ represent the same schedule but on permuted machines. In "symmetric" solutions, we have $\overline{i^s_{X_m}} = i^s_m$ $\forall m$, $t_{X_m}(\overline{\sigma}^s) = t_m(\sigma^s)$ $\forall m$ and $c(\overline{\sigma}^s) = c(\sigma^s)$. Then, $\overline{\sigma}^s$ is dominated by the "symmetric" solution $\sigma^s$ by Theorem 5.2.

Figure 5.6 illustrates Theorem 5.2, based on the example in Figure 2.8 (p. 36). Cost parameters are as follows: $w_{(i,j)} = p_{(i,j)}$ and $sc_{g,i} = 10st_{g,i}$. $\sigma^4$ and $\overline{\sigma}^4$ schedule the same set of jobs $\mathcal{AS}^4$, but the set of jobs is not the same on each machine $m$. For $X = (2, 1)$ we check $(i)$: $20 + 0 \leq 20$ $(m = 1)$ and $16 + 3 \leq 21$ $(m = 2)$. Checking $(ii)$, we have $(2 \cdot 7 + 30) - (0 + 30) \geq 10$, and $(iii)$ $18 \leq 22$ is also fulfilled, so that $\overline{\sigma}^4$ is dominated.

In B&B[$P\diamond ia\text{-}pb$] we store information for the dominance rule only for the job set $\mathcal{AS}^s$, and overwrite this information if the start times of a new $\sigma^s_{new}$ are greater or equal on all machines and if costs are less or equal than in $\sigma^s$. Therefore, it is important to store only "good" partial schedules because overwriting the information of a job set $\mathcal{AS}^s$ is then unlikely once an information is stored. This explains the importance of the branching rule described above.

**Computational results** for B&B[$P\diamond ia\text{-}pb$] are presented only for $\rho = $ M. CPU times increase (decrease) for $\rho = $ L (H) so that we do not vary $\rho$. We choose again a random setup structure $rd$ since in this case the setup matrix does *not* suggest partitioning of families for the different machines.[5] Again, results are only preliminary.

In order to test B&B[$P\diamond ia\text{-}pb$], we varied the problem size and the number of machines; we consider two problem classes $(M, N, J) = (2, 6, 24)$ and $(M, N, J) = (3, 6, 18)$ with 30 instances each. Instances are generated such that there is a feasible EDDWF schedule. For larger problem sizes with $(M, N, J) = (2, 6, 30)$ $((3, 6, 24))$, only 12 (1) out of 30 (30) instances are solved within the time limit of 1200 sec and the memory limit of 15 MB.[6] As for $\alpha = ML$ the rightshift-rule effectively curtails the enumeration. The dominance rule of Theorem 5.2 in B&B[$P\diamond ia\text{-}pb$] is not as effective as in the single-machine case, but more effective than for $\alpha = ML$, probably because of the more elaborated branching rule.

Table 5.5 displays the results. In contrast to the multi-level multi-machine case running times increase in the parallel machine case if the number of machines increases; since jobs can be assigned to different machines by $MA_k$, the enumeration tree grows with a larger $M$. Furthermore, only much smaller problem sizes compared to the single-machine

---

[5]E.g. the setup structure $gp$ partitions families into 3 setup groups, and one would straightforwardly solve 3 single-machine problems with 2 families instead. Therefore, a heuristic approach could be $(i)$ to cluster families based on an examination of the setup matrix and $(ii)$ to decompose the parallel machine problem into $M$ single-machine problems.

[6]If an instance could not be solved, B&B[$P\diamond ia\text{-}pb$] always exceeded the time limit of 1200 sec in our experiments.

**Table 5.5:** *Preliminary Computational Results for $\alpha = P$*

| | B&B[$P\diamond ia\text{-}pb$] | | | |
|---|---|---|---|---|
| | $(M, N, J) = (2, 6, 24)$ | | $(M, N, J) = (3, 6, 18)$ | |
| $\rho = M$ | $R_{avg}$ | $R_{max}$ | $R_{avg}$ | $R_{max}$ |
| | 108.5 | 568.3 | 143.1 | 935.2 |

(IBM PowerPC)

case (cf. Figure 3.4, p. 68) can be solved to optimality. New bounding rules (and better computational results) can be found in Kurth [78].

Recall, that we consider only problems with $N > M$; the case $N = M$ can be trivially solved by assigning each family to one machine.

# Chapter 6

# Artificial Intelligence Approaches

This chapter deals with artificial intelligence approaches for solving the BSP. We compare two solvers based on the artificial intelligence paradigm and a mixed-integer programming (MIP) solver with each other.

After an introduction in Section 6.1, we provide models and encodings for the different solvers in Section 6.2. In Section 6.3 some computational experience is reported. Conclusions follow in Section 6.4.

## 6.1   Introduction

General solvers are used if there is not enough time (or it is not worthwhile) to develop special purpose algorithms. In these cases the user is interested in an easily manageable tool to implement and solve his specific problem. A general solver has the advantage that no algorithm must be specified; only the problem must be represented in an appropriate form (which will be our concern in Section 6.2). From the user's point of view two aspects are of interest: the *modeling* aspect examines the question of how to "get the problem into the computer", i.e. whether we are able to encode constraints easily and to produce "readable" code which is easy to modify. The *performance* aspect concerns the running times of the different solvers.

We use the (state-of-the-art) MIP solver OSL provided by IBM (cf. IBM [69]) and the two constraint programming (CP) solvers CHARME (cf. BULL [23]) and clp(FD) (cf. Codognet and Diaz [29]). As an example problem we solve $[1/fam,ia\text{-}pb,st_{g,i},d_{(i,j)}/\gamma]$ for objectives $\gamma = \{\sum w_{(i,j)} C_{(i,j)}, \sum sc_{g,i}, *, \}$ with all solvers. A missing entry for $\gamma$ denotes the feasibility problem. *Specialized* algorithms which can solve instances with much larger problem sizes are presented in Section 3.2. As "information input" for all solvers we encode the different objectives $\gamma$, the sequencing constraints on the machine, and the service constraints (cf. also Section 2.3). Furthermore, we formulate the EDDWF ordering of jobs as constraints.[1] The information input represents the users knowledge about "what the problem is". But the user employs a general solver as he does not know "how to solve the problem" and consequently we do not implement any algorithmic (or problem specific) knowledge[2] in any solver.

A problem stated as a **MIP model** can be solved by any **MIP solver**. MIP solvers employ a linear programming (LP)-based branch and bound (B&B) algorithm; enhancing the performance of this algorithm is the major concern of all commercial vendors of MIP solvers.[3] As nonlinear expressions are not allowed, MIP models often have to comprise many decision variables which increase the running times of the underlying B&B algorithm. MIP models for scheduling problems are largely intractable and not easy to formulate in general. The solution times heavily depend on the quality of the LP-bounds, and the LP-bounds in turn depend on the structure of the MIP model.[4]

**CP solvers** are also denoted as CP languages; further terms are constraint (logic) programming, or finite domain languages or solvers. In the sequel we employ the term CP

---

[1]The model in Table 2.4 also allows schedules where jobs are not in EDDWF; it is the result of the *analysis* in Section 2.6 that only EDDWF schedules need to be considered.

[2]E.g. a feasibility bound as in Theorem 3.1 would represent algorithmic knowledge.

[3]There is also a broad range of research activities in this field, for recent developments cf. e.g. Mitra and Maros [91].

[4]A MIP solver works effectively if the MIP model generates "tight" LP-bounds. A lot of research (e.g. parts of polyhedral optimization) deals with the development of "tight" MIP models, cf. e.g. Eppen and Martin [47]. General statements on how to develop a "tight" MIP model systematically seem to be difficult to derive. As for specialized algorithms, the development of a "tight" MIP model is based on the analysis of the structural properties of the problem at hand.

solvers. An introduction to this field is given e.g. in Van Hentenryck [66]. CP solvers allow for a declarative, even a nonlinear problem description. Variables have an associated (finite) domain and constraints are used to reduce the size of the domains and to eliminate infeasible values. This value reduction is also referred to as *constraint propagation*. Constraint propagation is used during the enumeration to prune the search tree; essentially it involves consistency checking of domains.[5] If the domain of a variable consists of only one value, the variable is *instantiated* to that value. Local propagation mechanisms serve as bounds throughout the enumeration (in contrast to solutions of a relaxed subproblem, the LP-bounds of MIP solvers). Originally, CP solvers have been developed for constraint satisfaction problems where values for each variable (out of its domain) must be found so that all constraints are satisfied. Minimization of an objective function can be done by finding all solutions through enumeration and keeping the best objective function value as an upper bound. So we expect CP solvers to perform well on optimization problems with a *small* solution space. Different CP solvers differ not only in the syntax but also in the way how constraint propagation is implemented. Other CP solvers than the ones mentioned above are described in, e.g., Van Hentenryck [66].

In a review of **related work**, we did not find any comparison between MIP and CP solvers on the same problem. An exception holds for Dhar and Ranganathan [35] who compare modeling capabilities and performance of a MIP solver versus an expert system for one instance of a course scheduling problem. The performance and versatility of the CP solver CHIP for different combinatorial problems is analyzed in Aggoun and Beldiceanu [4] and Dincbas et al. [37]. Codognet and Diaz [29] give an introduction to clp(FD) and compare the performance of CHIP versus clp(FD) on some example problems. The conceptual model presented below is similar to the one developed by Woodruff and Spearman [125]. Modeling with CHARME is examined in Drexl and Jordan [42] and [43].

A survey of MIP models for scheduling problems is given in Blazewicz et al. [16]. The model MIP A is similar to the one developed by Sielken [114]. The model MIP B presented

---

[5]Constraint propagation may be illustrated with the following (very simple) example: consider the domain variables $Y$, $X$, $Y$ in $2..5$, $X$ in $1..4$, and the constraint $Y < X$. Constraint propagation then reduces the domains to $Y$ in $2..3$ and $X$ in $3..4$. For each $X$ ($Y$) there is always an $Y$ ($X$) so that the constraint is satisfied.

**Table 6.1:** *BSP Parameters with a Single Index*

| | | |
|---|---|---|
| $j$ | : | job index, $j = 1, \ldots, J$, with $0$ $(J+1)$ dummy start (end) job |
| $\mathcal{F}_i$ | : | set of jobs which belong to family $i$, $i = 1, \ldots, N$ |
| $f_j$ | : | family $i$ to which job $j$ belongs to, $f_0 = 0$ |
| $d_j$ | : | deadline of job $j$ |
| $p_j$ | : | processing time of job $j$ |
| $w_j = h_i p_j$ | : | earliness weight of job $j$ |
| $\mathsf{st}_{[j,l]}(\mathsf{sc}_{[j,l]})$ | : | setup time (costs) between job $j$ and job $l$, |
| | : | $j = 0, \ldots, J$, $l = 1, \ldots, J$ |
| | : | $\mathsf{st}_{[j,l]}(\mathsf{sc}_{[j,l]}) = \begin{cases} st_{0,i}(sc_{0,i}) & j = 0, \ l \in \mathcal{F}_i \\ 0 & j, l \in \mathcal{F}_i \\ st_{g,i}(sc_{g,i}) & j \in \mathcal{F}_g, \ l \in \mathcal{F}_i \end{cases}$ |
| $B$ | : | big number |

below has been proposed by Coleman [31].

# 6.2   Model Formulations

The following models are presented in such a way that the general solvers can access the attributes of a job with a single index $j$ or $l$. Therefore the definition of the BSP parameters given in Table 6.1 differs from the one in Table 2.1.[6] We access the job at position $k$ by $S_k$, and inversely identify by $R_j$ the position of job $j$. By $f_j$ we denote the family $i$ job $j$ belongs to, and the set of jobs belonging to family $i$ is denoted by $\mathcal{F}_i$. In $\mathcal{F}_i$, jobs are labeled in order of increasing deadlines, i.e. $\forall j, l \in \mathcal{F}_i : j < l \Leftrightarrow d_j < d_l$. For the MIP models we must be able to express the setup times and setup costs between jobs

---

[6]If jobs are ordered in EDDWF, we can access or identify a *job* by an *s*-partial sequence of *families*. Consequently, if jobs are indexed with $(i, j)$ we must employ either a backward or forward scheduling scheme according to an EDDWF precedence graph (cf. Section 3.1). But since we do not want to specify "how to solve the problem", we must be able to access a job by a single index.

**Table 6.2:** *Conceptual Model*

---

**Decision variables**

$C_{[j]}$ :  Completion time of job $j$

$S$ :  Sequence of all jobs, $S_k$ is the job at position $k$

$R$ :  Positions of all jobs, $R_{[j]}$ is the position of job $j$

---

$$\text{Minimize } Z_{BSP} = \sum_{k=1}^{J} w_{[k]}(d_{[k]} - C_{[k]}) + sc_{[f_{[S_{k-1}]}, f_{[S_k]}]} \tag{6.1}$$

subject to

$$S_{[R_k]} = k \qquad\qquad\qquad k = 1, \ldots, J \tag{6.2}$$

$$R_{[j]} < R_{[l]} \qquad\qquad\qquad j < l;\ j, l \in \mathcal{F}_i;\ i = 1, \ldots, N \tag{6.3}$$

$$C_{[k]} \leq d_{[k]} \qquad\qquad\qquad k = 1, \ldots, J \tag{6.4}$$

$$C_{[S_{k-1}]} + st_{[f_{[S_{k-1}]}, f_{[S_k]}]} + p_{[S_k]} \leq C_{[S_k]} \qquad\qquad k = 1, \ldots, J \tag{6.5}$$

$$C_{[0]} = 0;\ S_0 = 0;\ C_{[S_J]} = d_{[S_J]} \tag{6.6}$$

---

by matrices $st_{j,l}$ and $sc_{j,l}$, respectively; both are enlarged matrices derived from setups between families $g$ and $i$.

The **conceptual model** presented in Table 6.2 is essentially the same model as in Table 2.4 (p. 25); however, we index the variables with a single index and the EDDWF ordering is explicitly formulated as a constraint. We relate $S$ and $R$ in constraints (6.2) and express the EDDWF ordering in (6.3). Deadlines are enforced in constraints (6.4), and sequencing constraints in (6.5). The objective (6.1) is formulated for $\gamma = \sum w_{(i,j)} C_{(i,j)} + \sum sc_{g,i}$, and either the first or the second term or both are left out for the other objectives. The model can be concisely formulated by using decision variables as indexes[7] and is therefore called a conceptual model. It cannot be solved by MIP solvers because the variables and their coefficients are not yet identified in the model in Table 6.2.

---

[7]For a similar idea cf. Woodruff and Spearman [125].

Table 6.3 provides the complete CHARME code for the constraint satisfaction or feasibility problem (without objective) of the example in Figure 2.3 (p. 27). Encoding the conceptual model of Table 6.2 in CHARME is nearly a "one-to-one" implementation. The parameters of Figure 2.3 and the decision variables are stated in the first part. The declaration of S and the statement all_diff(S) enforces S to be a permutation of jobs. The statement generate S starts the search for finding a sequence which satisfies all constraints. The corresponding output in Table 6.3 gives the job sequence and ranges for the completion times. Jobs $[1, 2, 3, \ldots, 7]$ of the CHARME code correspond to jobs $(1,1), (1,2), (2,1), \ldots, (3,3)$ in Figure 6.1. $\sigma_{left}$ ($\sigma_{right}$) depicts the corresponding schedule if all jobs complete as early (late) as possible, e.g. job 1 (=job $(1,1)$) must complete in the interval $[2..3]$.

The encoding for clp(FD) in Table 6.4 uses Prolog predicates and lies somewhere in between a declarative modeling and a solution procedure. Parameters are *facts*, a schedule is generated with a *rule*. In Table 6.4 we define the predicates solve() and lessequal() ourselves whereas member() and delete() are *built-in* predicates. In clp(FD) we access a job by its tuple $(i, j)$ and the parameters are defined accordingly.[8] The schedule generated by clp(FD) is a *list* with entries $[[i, j], C_{(i,j)}]$.[9] During the search new entries are added to the list like in a forward scheduling scheme. For our example, a feasible schedule is generated by the query

```
solve([[[0,0],0]],[[1,1],[2,1],[3,1]],Schedule).
```

In Table 6.4, the first solve() predicate fails because the list of eligible jobs is not empty. In the second definition of the solve() predicate, the query initializes the completion time of the first (dummy) job [0,0] to zero and unifies the set of eligible jobs ELJobs with [[1,1],[2,1],[3,1]] at the beginning. The next job to schedule is chosen out of the set of eligible jobs. In this way we enforce the EDDWF ordering of jobs. The following lines

---

[8]E.g. with the query "d(1,2,X)." the variable X is unified with $d_{(1,2)}$.

[9]In Prolog, variables must start with an uppercase letter, also *lists* may be variables. In L = [] the variable L is the empty list, in L = [Head|Tail] the variable Head is unified with the first *element* of the list, and Tail with the *rest* of the list. The comma operator ',' represents the logical AND, and a period '.' terminates a predicate.

**Table 6.3:** *CHARME Code of the Example in Figure 2.3*

```
/****** parameters ******/
Hor = 0..21;                                 /* planning horizon **/
Fam = 3;   JJ  = 7;

D   = [0,   8, 21,  9, 20, 10, 16, 21]  :: [0..JJ] ; /* deadlines */
P   = [0,   1,  2,  1,  1,  1,  2,  1]  :: [0..JJ] ; /* processing times */
F   = [0,   1,  1,  2,  2,  3,  3,  3]  :: [0..JJ] ; /* family of the job */

FJ1 = [     1,  2                    ] ; /* jobs of family 1 **/
FJ2 = [             3,  4            ] ; /* jobs of family 2 **/
FJ3 = [                     5,  6,  7] ; /* jobs of family 3 **/

ST  = [1, 2, 1,
       0, 1, 1,
       0, 0, 1,
       1 ,2, 0] :: [0..Fam, 1..Fam] ;  /* setup time matrice */
/****** decision variables ******/
array  CT   :: [0..JJ] of  Hor;       /* completion times of each job */
array  S    :: [0..JJ] of  0..JJ;     /* job at position k */
array  R    :: [0..JJ] of  0..JJ;     /* position of job j */
/****** main program ******/
{
 /***** S is a permutation of jobs  ******/
  all_diff(S);
 /***** enforce EDD ordering within families ******/
  for K in 1..JJ do S[ R[  K  ] ] = K;
  for J in 2..2  do R[ FJ1[ J-1 ]] < R[ FJ1[ J ]]
  for J in 2..2  do R[ FJ2[ J-1 ]] < R[ FJ2[ J ]]
  for J in 2..3  do R[ FJ3[ J-1 ]] < R[ FJ3[ J ]]
 /***** enforce sequence on the machine and deadlines ******/
  for K in 1..JJ do {
  CT[ K ] <= D[ K ];
  CT[ S[ K - 1 ] ]  +  ST[ F[ S[ K -1 ] ] , F[ S[ K ] ] ]
  + P[ S[ K ] ]  <=   CT[ S[ K ] ];
  }
 /***** initialize ******/
  CT[ S[JJ] ] <= D[S[JJ]];
  CT[ 0 ] = 0;  S[ 0 ] = 0; R[ 0 ] = 0;
 /***** find a feasible sequence ******/
  generate S;
  print S; print CT;
}
```

```
Output:
S  = [0,1,2,3,4,5,6,7]::[0..7]
CT = [0,2..3,4..5,6..7,7..8,9..10,11..16,12..21]::[0..7]
```

**Table 6.4:** clp(FD) *Code of the Example in Figure 2.3*

---

```
solve(Schedule,[],Schedule).    % ELJobs = [] ==> solution found %

solve([[[F1,N1],Completion1]|Tail],ELJobs,Schedule):-
    member([F2,N2],ELJobs),  % [F2,N2] = eligible job %
    delete(ELJobs,[F2,N2],ELJobsDel),
    N1 is N2+1,
    k(F2,FamN),
    (N2 < FamN -> ELJobsNew=[[F2,N1]|ELJobsDel];ELJobsNew=ELJobsDel),
                        % new list of ELJobs is ELJobsNew %
    d(F2,N2,Deadline),p(F2,N2,Processing),s(F1,F2,Setup),
                        % get data %
    Completion2 in 0..Deadline,
    lessequal(Completion1,Completion2,Processing+Setup),% Constraint: %
                        % Completion1+Processing+Setup=<Completion2 %
    solve([[[F2,N2],Completion2],[[F1,N1],Completion1]|Tail],
                                        ELJobsNew,Schedule).
                        % process rest of the list %

lessequal(X,Y,C):-          % X+C #=< Y with %
    X in 0..max(Y)-C,       % "in" constraint %
    Y in min(X)+C..infinity.

% Parameters %
k(1,2). k(2,2). k(3,3). % k(family, number of jobs in family) %
% d(i,j,x) deadline,  p(i,j,x) processing time of (i,j) %
d(1,1, 8).  p(1,1,1). d(1,2,21).  p(1,2,2).
d(2,1, 9).  p(2,1,1). d(2,2,20).  p(2,2,1).
d(3,1,10).  p(3,1,1). d(3,2,16).  p(3,2,2). d(3,3,21).  p(3,3,1).
% s(g,i,x): x : setup time family g to family i %
s(0,1,1). s(0,2,2). s(0,3,1).
s(1,1,0). s(1,2,1). s(1,3,1).
s(2,1,0). s(2,2,0). s(2,3,1).
s(3,1,1). s(3,2,2). s(3,3,0).
```

---

```
Query:
solve([[[0,0],0]],[[1,1],[2,1],[3,1]],Schedule).
Output:
Schedule = [[[3,3],12..21],[[3,2],11..16],[[3,1],9..10],
[[2,2],7..8],[[2,1],6..7],[[1,2],4..5],[[1,1],2..3],[[0,0],0]] ?
(10 ms) yes
```

---

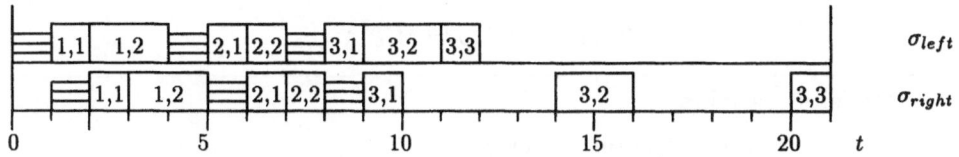

**Figure 6.1:** *Sequence and Ranges of Completion Times*

update ELJobs correctly. Afterwards, we formulate the constraints.[10] We then call the predicate solve() recursively to process the rest of the list, until all jobs are scheduled and the list of eligible jobs is empty.[11] The output of clp(FD) in Table 6.4 represents the same sequence and the same ranges of completion times as the output generated by CHARME (but in reverse order). The keyword "in" specifies the basic constraint of clp(FD); other constraints are defined with this basic constraint.[12]

MIP models of the BSP are less compact than the above conceptual model. In the constraints we have to consider explicitly that each job may be scheduled at (nearly) each position of the sequence. This leads to constraints for each pair $j, l$ of jobs. To achieve a better performance in our tests we employ different MIP models for different objectives $\gamma$.

Model **MIP A** in Table 6.5 is used for the objectives $\gamma \in \{\sum sc_{g,i}, *\}$. The binary decision variable $Y_{j,l}$ indicates whether job $l$ is sequenced consecutively after job $j$, which allows for the correct calculation of $sc_{j,l}$ in the objective (6.7). The variables $Y_{j,l}$ must define a permutation with 0 as the first, $J + 1$ as the last job, or, equivalently, each job must be scheduled exactly once before and behind another job, which is enforced by the (assignment) constraints (6.8) and (6.9) (cf. Blazewicz et al. [16]). For $B(1 - Y_{j,l}) = 0$,

---

[10]More precisely, member([F2,N2],ELJobs) chooses job [F2,N2] as the next job to schedule, and delete(ELJobs,[F2,N2],ELJobsDel) determines the jobs list ELJobsDel without [F2,N2]. If [F2,N2] is the last job of family N2, the deletion is correct, i.e. ELJobsNew=ELJobsDel, otherwise the next job of family N2 must be added so that ELJobsNew=[[F2,N1]|ELJobsDel]. Then, the attributes of [F2,N2] are determined and we can state the demand constraint Completion2 in 0..Deadline and the sequencing constraint lessequal(Completion1,Completion2,Processing+Setup).

[11]And then, for ELJobs=[], the first solve predicate succeeds.

[12]E.g., the definition of constraint lessequal() uses the keyword "in".

**Table 6.5:** *Model MIP A*

---

**Decision variables**

$C_j$      :      completion time of job $j$

$Y_{j,l}$    :    $\begin{cases} 1 & , & \text{if job } j \text{ is sequenced } consecutively \text{ before job } l \\ 0 & , & \text{otherwise} \end{cases}$

---

$$\text{Minimize } Z_{BSP} = \sum_{j=1}^{J} w_j(d_j - C_j) + \sum_{j=0}^{J} \sum_{\substack{l=1 \\ l\neq j}}^{J} \text{sc}_{j,l} Y_{j,l} \tag{6.7}$$

subject to

$$\sum_{\substack{l=1 \\ l\neq j;\ (j,l)\neq(0,J+1)}}^{J+1} Y_{j,l} = 1 \qquad\qquad j = 0,\ldots,J \tag{6.8}$$

$$\sum_{\substack{j=0 \\ j\neq l;\ (j,l)\neq(0,J+1)}}^{J} Y_{j,l} = 1 \qquad\qquad l = 1,\ldots,J+1 \tag{6.9}$$

$$C_j + \text{st}_{j,l} + p_l \leq C_l + B(1 - Y_{j,l}) \qquad \begin{cases} j = 0,\ldots,J;\ l = 1,\ldots,J+1; \\ j \neq l;\ (j,l) \neq (0,J+1) \end{cases} \tag{6.10}$$

$$C_j \leq d_j \qquad\qquad j = 1,\ldots,J \tag{6.11}$$

$$Y_{j,l} = 0 \qquad\qquad j > l;\ j,l \in \mathcal{F}_i;\ i = 1,\ldots,N \tag{6.12}$$

$$Y_{j,l} \in \{0;1\} \qquad\qquad j = 0,\ldots,J;\ l = 1,\ldots,J+1;\ j \neq l \tag{6.13}$$

$$C_0 = 0 \tag{6.14}$$

---

i.e. $j$ is consecutively scheduled before $l$, constraints (6.10) enforce the sequence on the machine. There are $J^2$ constraints (6.10) (for each possible pair of jobs) while there are only $J$ constraints (6.5) (only between consecutively sequenced jobs). Therefore, the MIP model needs a much larger number of sequencing constraints. Constraints (6.11) impose the deadlines. Due to the EDDWF ordering of jobs we know for two jobs $j, l \in \mathcal{F}_i$ with $j > l \Leftrightarrow d_j > d_l$ that $j$ is never scheduled before $l$ and hence we can fix a priori $Y_{j,l}$ to

**Table 6.6:** *Model MIP B*

---

**Decision variables**

$C_j$ : completion time of job $j$

$Y_{j,l}$ : $\begin{cases} 1 & , & \text{if job } j \text{ is sequenced before job } l \\ 0 & , & \text{otherwise} \end{cases}$

---

Minimize $Z_{BSP} = \sum_{j=1}^{J} w_j(d_j - C_j)$  $\hfill$ (6.15)

subject to

$\mathrm{st}_{0,j} + p_j \leq C_j \leq d_j$  $\hfill j = 0, \ldots, J$  $\hfill$ (6.16)

$C_j + \mathrm{st}_{j,l} + p_l \leq C_l + B(1 - Y_{j,l})$  $\hfill j = 1, \ldots, J;\ l = j+1, \ldots, J$  $\hfill$ (6.17)

$C_l + \mathrm{st}_{l,j} + p_j \leq C_j + BY_{j,l}$  $\hfill j = 1, \ldots, J;\ l = j+1, \ldots, J$  $\hfill$ (6.18)

$Y_{j,l} = 1$  $\hfill j < l;\ j, l \in \mathcal{F}_i;\ i = 1, \ldots, N$  $\hfill$ (6.19)

$Y_{j,l} \in \{0; 1\}$  $\hfill j = 0, \ldots, J;\ l = j+1, \ldots, J$  $\hfill$ (6.20)

---

$Y_{j,l} = 0$.

Model **MIP B** in Table 6.6 is used for the objective $\gamma = \{\sum w_{(i,j)} C_{(i,j)}\}$ as well as for the feasibility problem. $Y_{j,l} = 1$ denotes that job $l$ is scheduled (not necessarily consecutively) after job $j$. $Y_{j,l} = 0$ then denotes that $l$ is sequenced before $j$. We only need to consider $Y_{j,l}$ with $j < l$ (upper triangle) so that only half of the number of binary variables is needed (compared with the definition of $Y_{j,l}$ in MIP A). Constraints (6.16) define the time window for the completion time of job $j$. The constraints (6.17) and (6.18) express the disjunction and sequence the jobs on the machine; originally these constraints have been proposed by Baker [8]. Later on they have been extended by Coleman [31] for the case of sequence dependent setup times. Jobs are in EDDWF order, so some of the $Y_{j,l}$ can be fixed beforehand in (6.19) as done in MIP A.

Omitting the term with $\mathrm{sc}_{j,l}$ in MIP A would give rise to another MIP model for the objective $\gamma = \sum w_{(i,j)} C_{(i,j)}$, but the performance of MIP B is much better. Conversely, with numerous additional continuous variables, MIP B could also be extended to handle sequence dependent setup costs in the objective, but again the computational performance of MIP A is better despite the larger number of binary variables.

The models in Tables 6.5 and 6.6 can be implemented directly in a modeling language for MIP solvers. We use LINGO (cf. [87]) to generate the input for the MIP solver OSL.[13] Note that the (sometimes considerable long) generation time is not included in the CPU times given below.

## 6.3    Performance Comparison

All solvers employ some kind of generic B&B search. For a fair comparison we always apply the default enumeration strategy without any specific adaptation to the BSP. OSL employs a depth first search with some "built-in" enhancements specific to OSL. CHARME's enumeration strategy is to branch first to the variable with the smallest domain. For clp(FD) the branching scheme is specified by the way the predicate member() is applied. However, coding the BSP for clp(FD) this way, we employ a forward scheduling scheme, and one may argue that this is already a specific adaptation.

The CPU seconds in Table 6.7 are reported for OSL and CHARME running under AIX Unix on an IBM RS 6000/550 RISC workstation, and for clp(FD) running under PC Linux on a 486DX2/66 Mhz PC.[14] In Table 6.7 we vary the problem size $(N, J) = \{(3, 9), (3, 12)\}$ and the capacity utilization $\rho$=H, L, resulting in 4 problem classes. For each problem class 3 instances are generated by hand.[15] Table 6.7 reports the average time in seconds. Each instance is solved for the 4 objectives $\gamma = \{\sum w_{(i,j)} C_{(i,j)}, \sum sc_{g,i}, *, \}$, and with each of the

---

[13]LINGO generates so called MPS files. MPS is a standard format for the representation of the coefficients of a MIP model. MPS files can be used as input for any MIP solver.

[14]CPU times on the workstation and on the PC are in the same order of magnitude so that we do not use a correction factor for the slower PC.

[15]We generated the instances similar to the generator in Section 2.7. The 3 instances in each problem class have different setup structures: sequence independent $si$, sequence $sq$ with setup costs proportional to setup times and sequence $sq$ but arbitrary setup costs.

**Table 6.7:** *Performance Comparison between CHARME, clp(FD) and OSL*

| $(N,J)$ | $\rho$ | $\gamma = *$ | | | $\gamma = \sum sc_{g,i}$ | | |
|---|---|---|---|---|---|---|---|
| | | CHARME | clp(FD) | OSL (MIP A) | CHARME | clp(FD) | OSL (MIP A) |
| $(3,9)$ | H | 29.9 | 1.1 | 171.9 | 25.0 | 1.5 | 95.9 |
| | L | 73.1 | 14.6 | 33.2 | 47.0 | 15.9 | 7.1 |
| $(3,12)$ | H | 258.5 | 7.2 | **⁾871.4 | 234.3 | 7.3 | ***⁾ |
| | L | 1014.7 | 305.4 | 352.0 | 895.2 | 317.4 | 72.1 |

| $(N,J)$ | $\rho$ | $\gamma = \sum w_{(i,j)}C_{(i,j)}$ | | | $\gamma =$ | | |
|---|---|---|---|---|---|---|---|
| | | CHARME | clp(FD) | OSL (MIP B) | CHARME | clp(FD) | OSL (MIP B) |
| $(3,9)$ | H | 28.4 | 1.2 | 3.0 | 2.6 | 0.5 | 0.5 |
| | L | 51.9 | 14.3 | 8.2 | 1.1 | 0.1 | 0.4 |
| $(3,12)$ | H | 241.0 | 6.6 | 23.5 | 28.4 | 1.4 | 2.6 |
| | L | 822.2 | 299.5 | 124.1 | 8.8 | 0.2 | 2.6 |

**⁾: 2 of 3 problems not solved by OSL          (CHARME, OSL: RS 6000/550)

***⁾: 3 of 3 problems not solved by OSL               (clp(FD): 486DX2/66 PC)

3 solvers.

For the objectives $\gamma = *$ and $\gamma = \sum sc_{g,i}$, running times for CHARME and clp(FD) in Table 6.7 are small for $\rho$=H (small solution space, constraint propagation works effectively) but increase for $\rho$=L. In contrast, OSL is slow for $\rho$=H and becomes faster for $\rho$=L. Employing MIP A, OSL fails to solve 5 of 24 problems even in a CPU time of several hours.

If MIP B is used for $\gamma = \sum w_{(i,j)}C_{(i,j)}$, running times of OSL are faster compared to CHARME and faster compared to clp(FD) for $\rho$=L; the performance of the MIP solver highly depends on the model. Also, the running times of all solvers decrease for $\rho$=H. If only the constraint satisfaction problem is considered (no $\gamma$ entry), a solution is found more easily for $\rho$=L by all solvers (the solution space is larger). Running times are rather small compared to times for the optimization problems.

MIP solution times can be widespread for one problem class and objective, whereas solution times of CP solvers differ much less. Maximum solution times for high (low) capacity utilization are 1600 (543) sec for OSL, 401 (1070) sec for CHARME and 13 (412) sec for clp(FD).

For larger instances, the computational results of CHARME and OSL are discouraging. We tested the performance on instances where $(N, J) = (3, 18)$. OSL using MIP A did find no integer (feasible) solution at all. Only when using MIP B, integer solutions could be found within 50 sec. CHARME only finds feasible solutions (in about 100 sec) if all jobs are labeled in EDD order. clp(FD), however, only needs a few seconds to solve the feasibility problem, but cannot solve, e.g., instances with a problem sizes of $(N, J) = (5, 18)$.[16]

## 6.4   Conclusions

**Modeling** in CHARME can be done without having to define binary variables, via a conceptual model which uses decision variables as indexes. CHARME allows for a declarative modeling in equations with a syntax similar to the programming language C. clp(FD) has a Prolog-like syntax, and we must declare the problem with predicates, i.e. with rules and facts. But in the same way algorithmic knowledge can be implemented in terms of rules.[17] Prolog operates on *lists* which seem to be well suited to record the solutions of scheduling problems. The encoding of clp(FD) builds a list similar to the way $s$-partial schedules are generated by the B&B algorithm in Chapter 3, and this may be the reason for its good performance.

To achieve a comparable performance, for different objectives different MIP models are used — resulting in a higher modeling effort. Constraints must be linear so that we have much less freedom in the way how to model a problem. However, once the MIP model is built we can easily generate the input for a MIP solver with a modeling language like

---

[16]At least not with the current implementation. For the instances in Table 3.2 (p. 66), clp(FD) solves the feasibility problem in a maximum time of 175 sec for $(N, J) = (3, 25)$, but none of the $(N, J) = (5, 18)$ instances is solved – even in a CPU time of several hours.

[17]In Table 6.4, the construction of a schedule is formulated in terms of a rule, but in the same way we could include algorithmic knowledge, e.g. a feasibility bound.

LINGO. A general statement on which model provides the best "readability" and which is best suited to accommodate changes is difficult to derive. Moreover, this depends on the way of modeling with which a user is familiar.

With respect to the **performance** we observe that running times of the MIP solver OSL are highly unpredictable and that CP solvers seem to be more robust in this regard. High capacity utilization does not improve the LP-bounds for this sequencing problem, so that solution times of the MIP solver OSL sometimes are rather long. On the other hand, in the presence of high capacity utilization, constraint propagation allows for a more effective pruning of the search tree and hence makes CP solvers faster than the MIP solver in certain cases.[18]

An advantage of CP solvers is their ability to generate *all* solutions of a specific problem; constraints can be added or deleted during the search. Especially in an early stage of the modeling process this gives important insights (e.g. reveals unconsidered constraints) and is also well suited for multi-criteria decision making. Furthermore, partial solutions of the CP solver may be useful, unlike (partial) solutions of a MIP solver where some of the binary decision variables are not integer. Overall, clp(FD) seems to be the best general solver for the BSP. However, different results may hold for other problems.

---

[18]But also for CHARME this depends on the way a problem is modeled. Due to modeling freedom, each MIP model can also be implemented in CHARME (but not vice versa, see above). However, implementing a MIP model in CHARME, the LP-based B&B algorithm solves the problem more efficient. One has to find a concise model to employ CHARME effectively.

# Chapter 7

# Summary and Future Work

This final chapter summarizes the results of the work and points out directions for future research.

Batching and scheduling problems must be solved repeatedly in short term planning, which is the major motivation behind their treatment in this work.

Chapter 1 motivates the planning problem and introduces the batch sequencing problem (BSP). In Chapter 2 we propose a classification of research in the area of batching and scheduling, and we present models for several problem classes. In the analysis the BSP turns out to be NP-hard, but we managed to derive some structural properties of the BSP to restrict the search space for algorithms. An interesting result is that the concept of regenerative schedules is similar to a property for the uncapacitated single-product problem, found in 1958 by Wagner and Whitin (cf. [123]).

With respect to the methods for solving the BSP, described in Chapters 3 and 5 we were unable to develop a tight lower bound. An effective *exact* solution procedure is branch and bound search relying on a (dynamic programming-like) comparison of partial schedules, so that both time and memory requirements grow exponentially with problem size. However, up to now this procedure is still competitive when compared to other methods, but for practical applications the development of heuristics is required. A successful *heuristic* method tackles the batching and the scheduling decisions of the BSP separately, using problem specific knowledge for each one of the resulting subproblems.

Chapter 4 demonstrates that the well-known discrete lotsizing and scheduling problem (DLSP) solves the same planning problem as the BSP, based on the observation that DLSP and BSP generate corresponding solutions. Overall, it is easier to solve the planning problem in terms of a BSP because BSP decision variables "better" characterize a solution and hence make the methods to solve the BSP more effective.

In Chapter 5 we treat the multi-machine case; our contribution is the development of an exact algorithm. However, only instances with a small problem size could be solved to optimality.

Chapter 6 deals with constraint programming solvers based on the artificial intelligence paradigm. As to the BSP, it was found that modeling and solving with constraint programming solvers is – overall – superior to mixed-integer programming approaches.

Future work should focus on the fact that the BSP must be solved in a rolling horizon environment (where it is *a priori* known that only the first part of the solution will be implemented), and on situations, where parts of a schedule already exist.

Furthermore, this work leaves out the interaction with other planning levels such as medium term planning. An interesting approach is the one presented by Lasserre [81] and Dauzere-Peres and Lasserre [33] and [34] – an extension which included the BSP seems to be worthwhile.

# Appendix A

# Setup Matrices for the Instance Generator

All setup matrices $(st_{g,i})$ are given for $N = 6$ families

| Group ($gp$) | Random ($rd$) | Sequence ($sq$) | Sequence Independent ($si$) |
|:---:|:---:|:---:|:---:|
| 6 6 6 6 6 6 | 6 5 5 4 5 6 | 6 6 6 6 6 6 | 6 3 5 4 2 6 |
| 0 2 5 5 5 5 | 0 3 4 3 3 1 | 0 5 5 5 5 5 | 0 3 5 4 2 6 |
| 2 0 5 5 5 5 | 2 0 2 1 0 3 | 1 0 5 5 5 5 | 6 0 5 4 2 6 |
| 5 5 0 2 5 5 | 2 2 0 2 2 3 | 1 1 0 5 5 5 | 6 3 0 4 2 6 |
| 5 5 2 0 5 5 | 3 1 3 0 1 2 | 1 1 1 0 5 5 | 6 3 5 0 2 6 |
| 5 5 5 5 0 2 | 2 2 2 1 0 3 | 1 1 1 1 0 5 | 6 3 5 4 0 6 |
| 5 5 5 5 2 0 | 4 2 4 3 2 0 | 1 1 1 1 1 0 | 6 3 5 4 2 0 |

# List of Notation

**Chapter 2**

| | |
|---|---|
| $i$ | index of the family, $i = 1, ..., N$ |
| $n_i$ | number of jobs in family $i$ |
| $J$ | total number of jobs, $J = \sum_{i=1}^{N} n_i$ |
| $st_{g,i}$ | setup time from family $g$ to family $i$ |
| $sc_{g,i}$ | setup cost from family $g$ to family $i$ |
| $h_i$ | holding cost per unit of family $i$ |
| $(i,j)$ | denotes the $j$-th job of family $i$, $i = 1, ..., N$, $j = 1, ..., n_i$ |
| $p_{(i,j)}$ | processing time of the $j$-th job of family $i$ |
| $d_{(i,j)}$ | deadline of the $j$-th job of family $i$ |
| $w_{(i,j)}$ | earliness weight of the $j$-th job of family $i$ |
| | $w_{(i,j)} = h_i p_{(i,j)}$ |
| $\pi$ | sequence of all jobs |
| $k$ | index of the positions in $\pi$ |
| $(i_{[k]}, j_{[k]})$ | the job at position $k$ |
| $\sigma$ | schedule |
| $C_{(i,j)}$ | completion time of job $(i,j)$ |
| $F$ | flow-shop with identical machines |
| $ML1$ | multi-level, single-machine |
| $ML$ | multi-level, multi-machine, fixed family-machine assignment |
| $P$ | identical parallel machines |
| $fam$ | family scheduling problem |
| $\sum w_{(i,j)} C_{(i,j)}$ | objective: minimize earliness costs |

| | |
|---|---|
| $\sum sc_{g,i}$ | objective: minimize sequence dependent setup costs |
| $\sum sc_i$ | objective: minimize sequence independent setup costs |
| $*$ | $\sum w_{(i,j)} C_{(i,j)} + \sum sc_{g,i}$ or $\sum w_{(i,j)} C_{(i,j)} + \sum sc_i$ |
| $r_{(i,j)}$ | release date of the $j$-th job of family $i$ |
| $P_k$ | $\begin{cases} 1 & \text{if idle time preempts production between jobs} \\ & (i_{[k-1]}, j_{[k-1]}) \text{ and } (i_{[k]}, j_{[k]}) \\ 0 & \text{otherwise} \end{cases}$ |
| $\mathcal{E}$ | set of end families (multi-level models) |
| $\mathcal{I}$ | set of intermediate families (multi-level models) |
| $u_i$ | unique successor of family $i$, $i \in \mathcal{I}$ |
| $a_{i,u_i}$ | gozinto factor, number of units of $i$ needed per unit of $u_i$ |
| $m$ | machine index $m = 1, \ldots, M$ |
| $ma_i$ | machine $m$ on which family $i$ has to be produced |
| $\tilde{d}_{(i,j)}$ | multi-level deadline of $(i,j)$ |
| $z(k)$ | predecessor operator for multi-machine models |
| $MA_k$ | the machine $m$ to which the job at position $k$ is assigned (parallel machines) |
| $d^b_{(i,b)}$ $(w^b_{(i,b)})$ | deadline (earliness weight) of the $b$-th batch of family $i$ |
| $C^b_{(i,b)}$ | batch completion time |
| $gp$ | setup structure where families in $st_{g,i}$ form groups |
| $rd$ | setup structure where setups in $st_{g,i}$ are random |
| $sq$ | setup structure where setups in $st_{g,i}$ are large from $g$ to $i$, $g < i$ and small for $g > i$ |
| $\rho = H$ (M, L) | high (medium, low) capacity utilization |
| $\theta = h$ ($l$) | high (low) setup significance |

## Chapter 3

| | |
|---|---|
| $s$ | index of the (backward) scheduling stage |
| $\pi^s$ | $s$-partial sequence |
| $\sigma^s$ | $s$-partial schedule |

| | |
|---|---|
| $\overline{\sigma}^s$ | $s$-partial schedule currently under consideration, which may be dominated by the previously enumerated schedule $\sigma^s$, $\mathcal{AS}^s = \overline{\mathcal{AS}^s}$ |
| $\omega^s$ | completion of $\sigma^s$ |
| $(i^s, j^s)$ | job under consideration at stage $s$ |
| $t(\sigma^s)$ | start time of $\sigma^s$ |
| $c(\sigma^s)$ | costs of $\sigma^s$ |
| $\mathcal{AS}^s(\mathcal{US}^s)$ | set of jobs already scheduled (unscheduled) in the $s$-partial schedule $\sigma^s$ |
| $\mathcal{UI}^s$ | set of items to which jobs in $\mathcal{US}^s$ belong |
| $v_i$ | number of unscheduled jobs of family $i$ |
| $UB$ | upper bound, cost of the current best solution |
| $\xi_{t2c}$ | time cost tradeoff factor, $\xi_{t2c} = \sum_{(i,j) \in \mathcal{US}}^s w_{(i,j)}$ |
| $T^s$ $(C^s)$ | lower bound on the time (costs) to schedule a completion of $\sigma^s$ |
| $R_{avg}$ $(R_{max})$ | average (maximal) CPU time |
| $A_{avg}$ $(A_{max})$ | average (maximal) deviation in % from the optimal solution |
| $\#_{inf}$ | number of problems found infeasible |
| $Z_{BSP}(\sigma)$ | objective function value of $\sigma$ |
| $bc_{k_1, k_2}$ | block costs |
| $bs_k$ | block size |
| $gs_k$ | group size |
| $\mathcal{G}_1(\sigma^s)$ | first block of $\sigma^s$ |
| $w_1(\sigma^s)$ | earliness weights of jobs in $\mathcal{G}_1(\sigma^s)$ |
| $pbt(\sigma^s)$ | pull-back time |
| $p_{(i,b)}^b$ | batch processing time |
| $btl_{k_1, k_2}$ | batch length |
| $ls_k^*$ | latest start time of the job at position $k$ |
| $bts_k^*$ $(bts^s)$ | optimal batch size at position $k$ (of the first batch of $\sigma^s$) |

## Chapter 4

| | |
|---|---|
| $t$ | index of periods, $t = 1, \ldots, T$ |

| | |
|---|---|
| $q_{i,t}$ | demand of item $i$ in period $t$ |
| $sc^p_{g,i}$ | setup costs per setup period from item $g$ to item $i$ |
| $Y_{i,t}$ | 1, if item $i$ is produced in period $t$, and 0, otherwise |
| $V_{g,i,t}$ | 1, if the machine is setup for item $i$ in period $t$, while the previous item was item $g$, and 0, otherwise |
| $I_{i,t}$ | inventory of item $i$ at the end of period $t$ |
| $\nu$ | period-item assignment = solution for the DLSP |
| $\triangle_{avg}\ (\triangle_{max})$ | average (maximal) deviation in % from a lower bound |

## Chapter 5

| | |
|---|---|
| $m^s$ | the machine $m$ on which $(i^s, j^s)$ is scheduled |
| $i^s_m$ | the $s$-partial schedule $\sigma^s$ starts with family $i^s_m$ on machine $m$ |
| $t_m(\sigma^s)$ | start time of the $s$-partial schedule $\sigma^s$ on machine $m$ |
| $AS^s_m(US^s_m)$ | set of jobs already scheduled (unscheduled) on machine $m$ |
| $UI^s_m$ | set of families to which the jobs in $US^s_m$ belong |
| $X$ | matching vector, defining a permutation of machines |
| | $X = (X_1, \ldots, X_m, \ldots, X_M)$ |

## Chapter 6

| | |
|---|---|
| $j$ | job index, $j = 1, \ldots, J$, with $0\ (J+1)$ dummy start (end) job |
| $\mathcal{F}_i$ | set of jobs which belong to family $i$, $i = 1, \ldots, N$ |
| $f_j$ | family $i$ to which job $j$ belongs, $f_0 = 0$ |
| $d_j$ | deadline of job $j$ |
| $p_j$ | processing time of job $j$ |
| $w_j = h_i p_j$ | earliness weight of job $j$ |
| $\mathsf{st}_{[j,l]}(\mathsf{sc}_{[j,l]})$ | setup time (costs) between job $j$ and job $l$, $j = 0, \ldots, J$, $l = 1, \ldots, J$ |

$$\mathsf{st}_{[j,l]}(\mathsf{sc}_{[j,l]}) = \begin{cases} st_{0,i}(sc_{0,i}) & j = 0,\ l \in \mathcal{F}_i \\ 0 & j, l \in \mathcal{F}_i \\ st_{g,i}(sc_{g,i}) & j \in \mathcal{F}_g,\ l \in \mathcal{F}_i \end{cases}$$

$S$ permutation of jobs

$R_{[j]}$ position of job $j$

$C_{[j]}$ completion time of job $j$

# List of Abbreviations

| | |
|---|---|
| cf. | confer |
| e.g. | for example |
| i.e. | that is |
| w.r.t. | with respect to |
| w.l.o.g. | without loss of generality |

## Chapter 1

| | |
|---|---|
| BSP | batch sequencing problem |
| CLSP | capacitated lotsizing problem |
| ELSP | economic lot scheduling problem |
| DLSP | discrete lotsizing and scheduling problem |
| PPC | production planning and control |
| JIT | just-in-time |

## Chapter 2

| | |
|---|---|
| TSP | traveling salesman problem |
| $ia\text{-}pb$ | item availability-preemptive batching |
| $ia\text{-}npb$ | item availability-nonpreemptive batching |
| $ba$ | batch availability |
| $U$ | units |
| $MU$ | monetary units |

| | |
|---|---|
| $TU$ | time units |
| MIP | mixed-integer programming |
| EDD (ERD) | earliest deadline (release dates) ordering |
| SWPT | shortest weighted processing time ordering |
| EDDWF | earliest deadline within families |

## Chapter 3

| | |
|---|---|
| $[1 \diamond ia\text{-}pb]$ | concerning problem $[1/fam, ia\text{-}pb, st_{g,i}/\sum w_{(i,j)} C_{(i,j)} + \sum sc_{g,i}]$ |
| B&B | branch and bound algorithm |
| DP | dynamic programming algorithm |
| C&I | construction and improvement algorithm |
| TSWS | tabu search procedure of Woodruff and Spearman [125] |
| $[1 \diamond ia\text{-}npb]$ | concerning problem $[1/fam, ia\text{-}npb, st_{g,i}/\sum w_{(i,j)} C_{(i,j)} + \sum sc_{g,i}]$ |
| $[1 \diamond ia\text{-}npb, st_i]$ | concerning problem $[1/fam, ia\text{-}npb, st_i/\sum w_{(i,j)} C_{(i,j)} + \sum sc_i]$ (sequence independent setups) |
| GA | genetic algorithm |
| Phase II-Scheduling | algorithm to derive a schedule from a genetic string in the GA |
| $[1 \diamond ba]$ | concerning problem $[1/fam, ba, st_{g,i}/\sum w_{(i,j)} C_{(i,j)} + \sum sc_{g,i}]$ |
| $[F \diamond ba, st_i]$ | concerning problem $[F/fam, ba, st_i/\sum w_{(i,j)} C_{(i,j)} + \sum sc_i]$ |

## Chapter 4

| | |
|---|---|
| BSP(DLSP) | BSP instance derived from a DLSP instance |
| BSPUT(DLSP) | BSP instance with unit time jobs derived from a DLSP instance |
| SISTSC | DLSP with sequence independent setup times and setup costs |
| DACGP | dual ascent and column generation procedure to solve SISTSC |
| TSPTW | traveling salesman problem with time windows |
| SDSC | DLSP with sequence dependent setup costs and zero setup times |
| TSPOROPT | traveling salesman procedure and improvement |

|          | heuristic to solve SDSC                                           |
|----------|------------------------------------------------------------------|
| SDSTSC   | DLSP with sequence dependent setup times and setup costs         |
| TSPTWA   | DP algorithm to solve TSPTW, adapted to solve SDSTSC             |
| BA       | DLSP with batch availability                                     |
| BRJSA    | simulated annealing approach to solve BA                         |
| BA2LV    | DLSP with two levels and batch availability                     |
| BRJSA2LV | simulated annealing approach to solve BA2LV                      |

## Chapter 5

|                     |                                                                                              |
|---------------------|----------------------------------------------------------------------------------------------|
| LIN, GEN,           |                                                                                              |
| CON                 | gozinto graphs for the computational experiments                                             |
| SEP (INT)           | separated (interdependent) machine assignment                                                |
| $[ML1\diamond ia\text{-}pb]$ | concerning problem $[ML1/fam,st_{g,i},ia\text{-}pb/\sum w_{(i,j)}C_{(i,j)}+\sum sc_{g,i}]$   |
| $[ML\diamond ia\text{-}pb]$  | concerning problem $[ML/fam,st_{g,i},ia\text{-}pb/\sum w_{(i,j)}C_{(i,j)}+\sum sc_{g,i}]$    |
| $[P\diamond ia\text{-}pb]$   | concerning problem $[P/fam,st_{g,i},ia\text{-}pb/\sum w_{(i,j)}C_{(i,j)}+\sum sc_{g,i}]$     |

## Chapter 6

|                 |                                                  |
|-----------------|--------------------------------------------------|
| CP              | constraint programming                           |
| LP              | linear programming                               |
| OSL             | optimization subroutine library (MIP solver)     |
| CHARME,         |                                                  |
| clp(FD), CHIP   | CP solvers                                        |

# Bibliography

[1] AFENTAKIS, P. AND B. GAVISH, 1986. Optimal lotsizing algorithms for complex product structures. *Operations Research*, Vol. 34, pp. 237-249.

[2] AHMADI, R.H., S. DASU AND C. TANG, 1992. The dynamic line allocation problem. *Management Science*, Vol. 38, pp. 1341-1353.

[3] AHMADI, J.H., R.H. AHMADI, S. DASU AND C. TANG, 1992. Batching and scheduling jobs on batch and discrete processors. *Operations Research*, Vol. 39, pp. 750-763.

[4] AGGOUN, A. AND N. BELDICEANU, 1993. Extending CHIP in order to solve complex scheduling and placement problems. *Mathematical Computation and Modelling*, Vol. 17, pp. 57-73.

[5] AHN, B.H. AND J.H. HYUN, 1990. Single facility multi-class job scheduling. *Computers and Operations Research*, Vol. 17, pp. 265-272.

[6] ALBERS, S. AND P. BRUCKER, 1994. The complexity of one-machine batching problems. Working Paper, University of Osnabrück. To appear in *Discrete Applied Mathematics*.

[7] ARKIN, E.M. AND R.O. ROUNDY, 1991. Weighted-tardiness scheduling on parallel machines with proportional weights. *Operations Research*, Vol. 39, pp. 64-81.

[8] BAKER, K.R., 1974. *Introduction to Sequencing and Scheduling*, Wiley, New York.

[9] BAKER, K.R. AND L.E. SCHRAGE, 1978. Finding an optimal sequence by dynamic programming: an extension to precedence related tasks. *Operations Research*, Vol. 26, pp. 111-120.

[10] BARNES, J.W. AND L.K. VANSTON, 1977. Scheduling jobs with linear delay penalties and sequence dependent setup costs. *Operations Research*, Vol. 29, pp. 146-160.

[11] BEAN, J.C., 1994. Genetic algorithms and random keys for sequencing and optimization. *ORSA Journal on Computing*, Vol. 6, pp. 154-160.

[12] BEDNARZIG, J., 1994. *Das Job-shop Problem mit reihenfolgeabhängigen Rüstzeiten.* PhD-Thesis, University Hamburg, Germany.

[13] BERNARDO, J.J. AND K.S. LIN, 1994. An interactive procedure for bi-criteria production scheduling. *Computers and Operations Research*, Vol. 21, pp. 677-688.

[14] BIANCO, L. AND S. RICCIARDELLI, 1982. Scheduling of a single machine to minimize total weighted completion time subject to release dates. *Naval Research Logistics*, Vol. 29, pp. 151-167.

[15] BIANCO, L., S. RICCIARDELLI, G. RINALDI AND A. SASSANO, 1988. Scheduling tasks with sequence-dependent processing times. *Naval Research Logistics*, Vol. 35, pp. 177-184.

[16] BLAZEWICZ, J., M. DROR AND J. WEGLARZ, 1991. Mathematical programming formulations for machine scheduling: a survey. *European Journal of Operational Research*, Vol. 51, pp. 283-300.

[17] BRUCKER, P., M.Y. KOVALYOV AND F. WERNER, 1994. Parallel machine batch scheduling with deadlines and sequence independent set-ups. Working Paper, University of Osnabrück, Germany.

[18] BRÜGGEMANN, W., 1995. Ausgewählte Probleme der Produktionsplanung – Modellierung, Komplexität und neuere Lösungsmöglichkeiten. Production and Logistics, Physica-Verlag, Heidelberg, Germany.

[19] BRÜGGEMANN, W. AND H. JAHNKE, 1992. DLSP with multi item batch production. Conference proceedings, Production planning and control, Hagen, Germany, 1992, pp. 459-472.

[20] BRÜGGEMANN, W. AND H. JAHNKE, 1994. DLSP for 2-stage multi item batch production. *International Journal of Production Research*, Vol. 32, pp. 755-768.

[21] BRÜGGEMANN, W. AND H. JAHNKE, 1994. Remarks on: "Some extensions of the discrete lotsizing and scheduling problem" by Salomon, M., L.G. Kroon, R. Kuik and L.N. Van Wassenhove. Working Paper, Institut für Logistik and Transport, University Hamburg, Germany.

[22] BRUNO, J. AND P. DOWNEY, 1978. Complexity of task sequencing with deadlines, setup-times and changeover costs. *SIAM Journal on Computing*, Vol. 7, pp. 393-404.

[23] BULL S.A., 1990. *Charme VI User's Guide and reference manual.* Artificial intelligence development centre. BULL S.A., Paris, France.

[24] CATTRYSSE, D., M. SALOMON, R. KUIK AND L.N. VAN WASSENHOVE, 1993. A dual ascent and column generation heuristic for the discrete lotsizing and scheduling problem with setup-times. *Management Science*, Vol. 39, pp. 477-486.

[25] CATTRYSSE, D. AND J. MAES, 1990. Set partitioning and column generation heuristics for capacitated dynamic lotsizing. *European Journal of Operational Research*, Vol. 46, pp. 38-47.

[26] CHANDRU, V., C.Y. LEE AND R. UZSOY, 1993. Minimizing total completion time on a batch processing machine with job families. *Operations Research Letters*, Vol. 13, pp. 61-65.

[27] CHANDRU, V., C.Y. LEE AND R. UZSOY, 1993. Minimizing total completion time on batch processing machines. *International Journal of Production Research*, Vol. 31, pp. 2097-2121.

[28] CHENG, T.C.E., T.-L. CHEN AND C. OGUZ, 1994. One machine batching and sequencing of multiple-type items. *Computers and Operations Research*, Vol. 21, pp. 717-721.

[29] CODOGNET, P. AND D. DIAZ, 1994. Compiling constraints in clp(FD). *Journal of Logic Programming*, Vol. 19, pp. 1-41.

[30] COFFMAN, E.G.JR., M. YANNAKAKIS, M. MAGAZINE AND C. SANTOS, 1990. Batch sizing and job sequencing on a single machine. *Annals of Operations Research*, Vol. 26, pp. 135-147.

[31] COLEMAN, B.J., 1992. A simple model for optimizing the single machine early/tardy problem with sequence dependent setups. *Production and Operations Management*, Vol. 1, pp. 225-228.

[32] CONWAY, R.W., W.L. MAXWELL AND L.W. MILLER, 1967. *Theory of scheduling*, Addison-Wesley, Massachusetts, USA.

[33] DAUZERE-PERES, S. AND J.B. LASSERRE, 1994. *An integrated approach in production planning and scheduling*. Lecture Notes in Economics and Mathematical Systems No. 411, Springer-Verlag, Berlin, Germany.

[34] DAUZERE-PERES, S. AND J.B. LASSERRE, 1994. Integration of lot-sizing and scheduling decisions in a job-shop. *European Journal of Operational Research*, Vol. 75, pp. 413-426.

[35] DHAR, V. AND N. RANGANATHAN, 1990. Integer programming versus expert systems. *Communications of the ACM*, Vol. 33, pp. 323-336.

[36] DIABY, M., H.C. BAHL, M.H. KARWAN AND S. ZIONTS, 1992. A lagrangean relaxation approach for very large scale capacitated lotsizing. *Management Science*, Vol. 38, pp. 1329-1340.

[37] DINCBAS, M., H. SIMONIS AND P. VAN HENTENRYCK, 1990. Solving large combinatorial problems in logic programming. *Journal of Logic Programming*, Vol. 8, pp. 75-93.

[38] DOBSON, G., 1992. The cyclic lot scheduling problem with sequence dependent setups. *Operations Research*, Vol. 40, pp. 736-749.

[39] DOMSCHKE, W., A. SCHOLL AND S. VOSS, 1993. *Produktionsplanung – Ablauforganisatorische Aspekte*. Springer-Verlag, Berlin, Germany.

[40] DREXL, A., K. HAASE AND A. KIMMS, 1995. Losgrößen– und Ablaufplanung in PPS–Systemen auf der Basis randomisierter Opportunitätskosten. *Zeitschrift für Betriebswirtschaft*, Jg. 65, pp. 267-285.

[41] DREXL, A., B. FLEISCHMANN, H.-O. GÜNTHER, H. STADTLER AND H. TEMPELMEIER, 1994. Konzeptionelle Grundlagen kapazitätsorientierter PPS–Systeme. *Zeitschrift für betriebswirtschaftliche Forschung*, Jg. 46, pp. 1022-1045.

[42] DREXL, A. AND C. JORDAN, 1994. Wissensbasierte Produktionsprozeßplanung, Planung bei reihenfolgeabhängigen Rüstkosten und -zeiten. *Zeitschrift für wirtschaftliche Fertigung*, 3/94, pp. 119-121.

[43] DREXL, A. AND C. JORDAN, 1995. Materialflußorientierte Produktionssteuerung bei Variantenfließfertigung. To appear in *Zeitschrift für betriebswirtschaftliche Forschung*.

[44] DRISCOLL, W.C. AND H. EMMONS, 1977. Scheduling production on one machine with changeover costs. *AIIE Transactions*, Vol. 9, pp. 388-395.

[45] DUMAS, Y., J. DESROSIERS, E. GELINAS AND M.M. SOLOMON, 1995. Technical Note: An optimal algorithm for the traveling salesman problem with time windows. *Operations Research*, Vol. 43, pp. 367-371.

[46] EMMONS, H., 1975. One machine sequencing to minimize mean flow time with minimum number tardy. *Naval Research Logistics Quarterly*, Vol. 22, pp. 585-592.

[47] EPPEN, G.D. AND R.K. MARTIN, 1987. Solving multi-item capacitated lot-sizing problems using variable redefinition. *Operations Research*, Vol. 35, pp. 587-598.

[48] ERLENKOTTER, D., 1990. Ford Whitman Harris and the economic order quantity model. *Operations Research*, Vol. 38, pp. 937-946.

[49] FLEISCHMANN, B., 1990. The discrete lot-sizing and scheduling problem. *European Journal of Operational Research*, Vol. 44, pp. 337-348.

[50] FLEISCHMANN, B., 1994. The discrete lot-sizing and scheduling problem with sequence-dependent setup-costs. *European Journal of Operational Research*, Vol. 75, pp. 395-404.

[51] FRENCH, S., 1982. *Sequencing and scheduling*. Ellis Horwood, Chichester, USA.

[52] GAREY, M.R. AND D.S. JOHNSON, 1979. *Computers and intractability - a guide to the theory of NP–completeness*. Freeman, San Francisco.

[53] GARFINKEL, R.S. AND G.L. NEMHAUSER, 1969. Set partitioning problem: set covering with equality constraints. *Operations Research*, Vol. 17, pp. 848-856.

[54] GASCON, A. AND R. LEACHMAN, 1988. A dynamic programming solution to the dynamic, multi item, single machine scheduling problem. *Operations Research*, Vol. 36, pp. 50-56.

[55] GHOSH, J.B., 1994. Batch scheduling to minimize total completion time. *Operations Research Letters*, Vol. 16, pp. 271-275.

[56] GLASSEY, C.R., 1968. Minimum changeover cost scheduling of several products on one machine. *Operations Research*, Vol. 16, pp. 343-352.

[57] GOLDBERG, D., 1989. *Genetic algorithms in search, optimization and machine learning*, Addison-Wesley, Reading, USA.

[58] GUPTA, S.K., 1982. $N$ jobs and $M$ machine job shop problems with sequence dependent setup times. *International Journal of Production Research*, Vol. 20, pp. 643-656.

[59] GUPTA, J.N.D., 1988. Single facility scheduling with multiple job classes. *European Journal of Operational Research*, Vol. 38, pp. 42-45.

[60] GUPTA, S.K. AND J. KYPARISIS, 1987. Single machine scheduling research. *OMEGA*, Vol. 15, pp. 207-227.

[61] HAASE, K., 1994. *Lotsizing and scheduling for production planning.* Lecture Notes in Economics and Mathematical Systems No. 408, Springer-Verlag, Berlin, Germany.

[62] HAASE, K., 1993. *Capacitated lot-sizing with linked production quantities of adjacent periods.* Working Paper No. 334, University of Kiel, Germany.

[63] HAASE, K. AND A. KIMMS, 1995. *Integrated lotsizing and scheduling with sequence dependent setup times.* Paper presented at the SOR September 1995, Passau, Germany.

[64] HAASE, K. AND L. GÖPFERT, 1995. *Engpaßorientierte Fertigungssteuerung bei reihenfolgeabhängigen Rüstvorgängen in einem Unternehmen der Satz– und Drucktechnik.* Working Paper No. 373, University of Kiel, Germany.

[65] HARIRI, A.M.A. AND C.N. POTTS, 1995. Single machine scheduling with deadlines to minimize the weighted number of tardy jobs. *Management Science*, Vol. 40, pp. 1712-1719.

[66] VAN HENTENRYCK, P., 1989. *Constraint satisfaction in logic programming.* The MIT Press, Cambridge/MA, USA.

[67] HOCHBAUM, D.S. AND D. LANDY, 1994. Scheduling with batching: minimizing the weighted number of tardy jobs. *Operations Research Letters*, Vol. 16, pp. 79-86.

[68] HU, T.C., Y.S. KUO AND F. RUSKEY, 1987. Some optimum algorithms for scheduling problems with changeover costs. *Operations Research*, Vol. 35, pp. 94-99.

[69] IBM CORPORATION, 1992. *OSL (Optimization Subroutine Library), guide and reference, release 2.* Kingston/NY, USA.

[70] JORDAN, C., 1995. A two phase genetic algorithm to solve variants of the batch sequencing problem. Working Paper No. 363, University of Kiel, Germany.

[71] JORDAN, C. AND A. DREXL, 1994. Lotsizing and scheduling by batch sequencing. Working Paper No. 343, University of Kiel, Germany.

[72] JORDAN, C. AND A. DREXL, 1995. A comparison of constraint and mixed-integer programming solvers for batch sequencing with sequence-dependent setups. *ORSA Journal on Computing*, Vol. 7, pp. 160-165.

[73] JULIEN, F.M. AND M. MAGAZINE, 1990. Scheduling customer orders, an alternative scheduling approach. *Journal of Manufacturing and Operations Management*, Vol. 3, pp. 177-199.

[74] TEN KATE, H.A., 1995. *Order acceptance and production control.* Theses on Systems, Organisations and Management, University of Groningen. Labyrint publication, Capelle aan den Ijssel, The Netherlands.

[75] KIMMS, A., 1993. Multi-level, single-machine lot sizing and scheduling (with initial inventory). To appear in *European Journal of Operational Research*.

[76] KOPPELMANN, J., 1995. *Ein exaktes Verfahren für das Multilevel Batch Sequencing Problem.* Diploma Thesis, University of Kiel, Germany.

[77] KUIK, R., M. SALOMON AND L.N. VAN WASSENHOVE, 1994. Batching decisions: structure and models. *European Journal of Operational Research*, Vol. 75, pp. 243-263.

[78] KURTH, J., 1996. *Batch sequencing problem – methods for the the parallel machine case.* Diploma Thesis, University of Kiel, Germany.

[79] LAGUNA, M., J.W. BARNES AND F. GLOVER, 1993. Intelligent scheduling with tabu search: an application to jobs with linear delay penalties and sequence dependent setup cost and times. *Journal of Applied Intelligence*, Vol. 3, pp. 159-172.

[80] LASDON L.S. AND R.C. TERJUNG, 1971. An efficient algorithm for multi-item scheduling. *Operations Research*, Vol. 19, pp. 946-969.

[81] LASSERRE, J.B., 1992. An integrated model for job-shop planning and scheduling. *Management Science*, Vol. 38, pp. 1201-1211.

[82] LEACHMAN, R.C., A. GASCON AND Z.K. XIONG, 1993. Multi-item single-machine scheduling with material supply constraints. *Journal of the Operational Research Society*, Vol. 44, pp. 1145-1154.

[83] LEE, C.Y., R. UZSOY AND L.A. MARTIN-VEGA, 1992. Efficient algorithms for scheduling semiconductor burn-in operations. *Operations Research*, Vol. 40, pp. 764-774.

[84] LEON, V.J. AND S.D. WU, 1992. On scheduling with ready times, due dates and vacations. *Naval Research Logistics*, Vol. 39, pp. 53-65.

[85] LIEPINS, G.E. AND M.R. HILLIARD, 1989. Genetic algorithms: Foundations and applications. *Annals of Operations Research*, Vol. 21, pp. 31-58.

[86] LIAO, C.J. AND L.M. LIAO, 1993. Minimizing the range of order completion times with multiple job classes. *Journal of the Operational Research Society*, Vol. 44, pp. 991-1002.

[87] LINDO SYSTEMS INC., 1993. *Lingo, optimization modeling language, guide and reference*. Chicago/IL, USA.

[88] LOCKET, A.G. AND A.P. MUHLMANN, 1972. Technical note: A scheduling problem involving sequence dependent changeover times. *Operations Research*, Vol. 20, pp. 895-902.

[89] MAGNANTI, T.L. AND R. VACCHANI, 1990. A strong cutting plane algorithm for production scheduling with changeover costs. *Operations Research*, Vol. 38, pp. 456-473.

[90] MASON, A.J. AND E.J. ANDERSON, 1991. Minimizing flow time on a single machine with job classes and setup times. *Naval Research Logistics*, Vol. 38, pp. 333-350.

[91] MITRA, G. AND I. MAROS (EDITORS), 1993. Applied mathematical programming and modelling. *Annals of Operations Research*, Vol. 43. J.C. Baltzer, Basel, Switzerland.

[92] MONMA, C.L. AND C.N. POTTS, 1989. On the complexity of scheduling with batch setup-times. *Operations Research*, Vol. 37, pp. 798-804.

[93] MONMA, C.L. AND C.N. POTTS, 1993. Analysis of heuristics for preemptive parallel machine scheduling with batch setup-times. *Operations Research*, Vol. 41, pp. 981-993.

[94] MOORE, J.E., 1975. An algorithm for a single machine scheduling problem with sequence dependent setup times and scheduling windows. *AIIE Transactions*, Vol. 7, pp. 35-41.

[95] NADDEF, D AND C. SANTOS, 1988. One-pass batching algorithms for the one-machine problem. *Discrete Applied Mathematics*, Vol. 21, pp. 133-145.

[96] OVACIC, I.M. AND R. UZSOY, 1993. Worst-case error bounds for parallel machine scheduling problems with bounded sequence dependent setup times. *Operations Research Letters*, Vol. 14, pp. 251-256.

[97] OVACIC, I.M. AND R. UZSOY, 1994. Rolling horizon algorithms for a single machine dynamic scheduling problem with sequence dependent setup times. *International Journal of Production Research*, Vol. 32, pp. 1243-1263.

[98] PARK, M., R. DATTERO AND J.J. KANET, 1993. Single machine batch scheduling with setup-times. Working Paper, Florida Atlantic University, USA.

[99] PINEDO, M., 1995. *Scheduling, theory, algorithms and systems*. Prentice-Hall, Englewood Cliffs/New Jersey, USA.

[100] POPP, T., 1993. *Das DLSP mit reihenfolgeabhängigen Rüstkosten*, Verlag Kovac, Hamburg, Germany.

[101] POTTS, C.N., 1991. Scheduling two classes of jobs on a single machine. *Computers and Operations Research*, Vol. 18, pp. 411-415.

[102] POTTS, C.N. AND L.N. VAN WASSENHOVE, 1992. Integrating scheduling with batching and lot-sizing: a review of algorithms and complexity. *Journal of the Operational Research Society*, Vol. 43, pp. 395-406.

[103] POSNER, M.E., 1985. Minimizing weighted completion times with deadlines. *Operations Research*, Vol. 33, pp. 562-574.

[104] POSNER, M.E., 1988. The deadline constrained weighted completion time problem: analysis of a heuristic. *Operations Research*, Vol. 36, pp. 778-782.

[105] PSARAFTIS, H.N., 1980. A dynamic programming approach for sequencing groups of identical jobs. *Operations Research*, Vol. 28, pp. 1347-1359.

[106] RAMAN, N. AND F.B. TALBOT, 1993. The job shop tardiness problem: A decomposition approach. *European Journal of Operations Research*, Vol. 69, pp. 187-199.

[107] RUBIN, P.A. AND G.L. RAGATZ, 1995. Scheduling in a sequence dependent setup environment with genetic search. *Computers and Operations Research*, Vol. 22, pp. 85-99.

[108] SALOMON, M., 1991. *Deterministic lotsizing models for production planning.* Lecture Notes in Economics and Mathematical Systems No. 355, Springer-Verlag, Berlin, Germany.

[109] SALOMON, M., L.G. KROON, R. KUIK AND L.N. VAN WASSENHOVE, 1991. Some extensions of the discrete lotsizing and scheduling problem. *Management Science*, Vol. 37, pp. 801-812.

[110] SALOMON, M., M.M. SOLOMON, L.N. VAN WASSENHOVE, Y.D. DUMAS, S. DAUZERE-PERES, 1995. Discrete lotsizing and scheduling with sequence dependent setup times and costs. Working Paper. To appear in *European Journal of Operational Research*.

[111] SANTOS, C. AND M. MAGAZINE, 1985. Batching in single operation manufacturing systems. *Operations Research Letters*, Vol. 4, pp. 99-103

[112] SCHUTTEN, J.M., S.L. VAN DE VELDE AND W.H. ZIJM, 1994. Single-machine scheduling with release dates, due dates and family setup times. To appear in *Management Science*.

[113] SCHUTTEN, J.M. AND R.A. LEUSSINK, 1995. Parallel machine scheduling with release dates, due dates and family setup times. To appear in *International Journal of Production Economics*.

[114] SIELKEN, R.L.JR., 1976. Sequencing with setup costs by zero-one mixed integer linear programming. *AIIE Transactions*, Vol. 8, pp. 369-371.

[115] SOTSKOV, Y.N., TAUTENHAHN, T. AND F. WERNER, 1994. Heuristics for permutation flow shop scheduling with batch setup times. Working Paper, University of Magdeburg, Magdeburg, Germany.

[116] SPRECHER, A., 1994. *Resource-constrained project scheduling - exact methods for the multi-mode case.* Lecture Notes in Economics and Mathematical Systems No. 409, Springer-Verlag, Berlin, Germany.

[117] SZWARC, W. AND J.J. LIU, 1993. Weighted tardiness single machine scheduling with proportional weights. *Management Science*, Vol. 39, pp. 626-632.

[118] TANG, C.S., 1990. Scheduling batches on parallel machines with major and minor setups. *European Journal of Operational Research*, Vol. 46, pp. 28-37.

[119] TEMPELMEIER, H. AND M. DERSTROFF, 1995. A lagrangean-based heuristic for dynamic multi-level multi-item constrained lotsizing with setup times. To appear in *Management Science*.

[120] TEMPELMEIER, H. AND S. HELBER, 1994. A heuristic for dynamic multi-item multi-level capacitated lotsizing for general product structures. *European Journal of Operational Research*, Vol. 75, pp. 296-311.

[121] UNAL, A. AND A.S. KIRAN, 1992. Batch sequencing, *IIE Transactions*, Vol. 24, pp. 73-83.

[122] VICKSON, R.G., M. MAGAZINE AND C. SANTOS, 1993. Batching and sequencing of components at a single facility. *IIE Transactions*, Vol. 25, pp. 65-70.

[123] WAGNER, H.M. AND T.M. WHITIN, 1958. Dynamic version of the economic lot size model. *Management Science*, Vol. 5, pp. 89-96.

[124] WEBSTER, S. AND K.R. BAKER, 1995. Scheduling groups of jobs on a single machine. *Operations Research*, Vol. 43, pp. 692-704.

[125] WOODRUFF, D.L. AND M.L. SPEARMAN, 1992. Sequencing and batching for two classes of jobs with deadlines and setup-times. *Production and Operations Management*, Vol. 1, pp. 87-102.

# List of Figures

# List of Tables

# Index

# Lecture Notes in Economics and Mathematical Systems

For information about Vols. 1–257
please contact your bookseller or Springer-Verlag